Meathead

UNRAVELING THE ATHLETIC BRAIN

Allison Brager

WestBow Press books may be ordered through booksellers or by contacting:

WestBow Press
A Division of Thomas Nelson & Zondervan
1663 Liberty Drive
Bloomington, IN 47403
www.westbowpress.com
1 (866) 928-1240

ISBN: 978-1-4908-6444-0 (sc)
ISBN: 978-1-4908-6445-7 (hc)
ISBN: 978-1-4908-6443-3 (e)

Library of Congress Control Number: 2015900496

Printed in the United States of America.

WestBow Press rev. date: 02/04/2015

Contents

Dedication

This book is dedicated to the coaches of my high school, college, and post-collegiate years for pushing me to my mental and physiological limits. Much gratitude is owed to two female coaches--Denise Gorski and Anne Rothenberg--who have superbly excelled in the androgynous world of sport. Mrs. G, words cannot thank you enough for the countless hours that you have invested in my success on and off the track and field. You taught me the importance of hard work, discipline, and perseverance. But it wasn't just me. You led many young, talented women into promising careers in track and field and life, as Adriane Blewitt-Wilson will touch upon in the foreword. I will never forget the emphasis that you placed on making eye contact when speaking or listening. It has stuck with me throughout my professional life. Anne, thank you for being mom away from home. Your emphasis on teamwork, excellence, and sacrifice has also carried over into my professional life. I must also thank my parents. I cannot count how many times my athletic aspirations (and injuries) have worried them, but yet their words and actions of encouragement have meant the world. Lastly, to my husband who constantly has to spend too much of our relationship at the gym.

Foreword

The career of the athlete is dependent on natural physical ability, but the guidance of a knowledgeable mentor and coach will make the difference in the athlete's overall performance. My experience as an athlete started at a young age when sport was disguised simply as playtime. I wasn't aware of physiologic processes, importance of diet, or how to set up a training cycle. I only knew how to move my body to hit a golf ball or throw a softball. As my body matured, the interest in sport also flourished. My development was in need of guidance as I stood at the crossroads of being an average high school athlete or a skilled track and field specialist with a promising future.

Denise Gorski was the head track and field coach at my alma mater, Boardman High School in northeast Ohio. Qualified with 35 years experience, "Mrs. G" celebrated the success of her tremendous female athletes at the state and national level. She was very active within many coaches' associations and went out of her way to continue her personal education as a coach. Mrs. G often stepped out of her comfort zone with the distance runners and worked with the jumpers, hurdlers, and throwers. Her willingness to learn and placing the priority on the athletes' needs produced multiple state qualifiers, all state placers and winners. The talent of the team was cultivated with her love and dedication, which in turn brought the team together as conference champions and lifelong friends after graduation.

Mrs. G mentored me in ways more than just an inspiring teacher, disciplined coach, or successful female professional. Mrs. G stood by me when our family suffered the tragic loss of my father at the start of my senior year at BHS; she helped me narrow down my path for collegiate athletics, and continued to rally for my success when cancer threatened my Olympic dreams. As an educator, Mrs. G emphasized the importance of accountability and hard work to all her students, even to the kids not involved in varsity sports. She nurtured my interest in sport and exercise when the majority of the girls in my class hated physical education and sweating in the beginning of the school day. Mrs. G gained the respect of the students and her colleagues with her disciplined attitude and mutual respect for others.

My relationship with Mrs. G is very special to me, but I am confident she has been the driving force for so many young female athletes. It is rare to find a coach who will take on the responsibility of an entire team of adolescent girls and still have the energy to guide them to athletic victories and personal growth. Denise Gorski is a tough leader and will fight for her athletes. In the personal case of the author, the inaugural year of girls' pole vault in the state of Ohio in 2002 was full of challenges and Allison was the best representative from BHS for the girls' competition. The state of Ohio was hesitant to allow girls to pole vault for a number of reasons. First, there would be an uneven number of events between the boys and girls. More importantly, there was concern over the growing statistics of injuries and deaths among high school boy vaulters, making pole vault be deemed too dangerous for girls. Mrs. G felt the sexist stand on the girls' event was wrong and was a strong voice to encourage Allison and the girls to compete.

At the end of my senior year, I qualified for the state track meet in the shot put and discus throw. I competed well in the first three rounds of the shot put and according to Mrs. G's score sheet, I had thrown far enough to earn three more throws in the finals. Unfortunately, the head official at the event had mismarked a throw on his sheet, which bumped me out of the competition. Mrs. G consulted with my Mom and other coaches' notes to confirm that the head official was incorrect. When she approached the official on my behalf, he would not budge on his ruling. The distances on his sheet would stand and I would not continue in the finals. Willing to put up a fight and defend my efforts, Mrs. G ran across the field into the track stadium at The Ohio State University and filed a protest against the official's ruling. In the meantime the event continued without me. I had a number of my competitors' coaches approach me with their notes and distances confirming I should be in the finals. All I could do was sit in the stands and cry about the mistake. The event ended and my high school shot put career was concluded. Moments later, Mrs. G ran up with the news that the meet committee evaluated her protest and granted me three more throws which would count towards the final results. Elated, I ran around simulating some sort of warm up and took three throws all by myself. I received an encouraging applause from the crowd when my best effort actually moved me up to sixth place and on the podium with the best athletes and the All-Ohio status.

As I recall that day in Columbus, I am reminded how dedicated Mrs. G was to me and to every single athlete on her team. She always went the extra mile to maximize any opportunity we had to achieve success, even after high school. I continued throwing in college at Ashland University and had an excellent season in 2003 leading up to the 2004 US Olympic Trials. Before I had the chance to compete for a spot on the US Olympic Team, I was diagnosed with Hodgkin's Lymphoma, a rare cancer among young people. I endured six months of chemotherapy and even though I had a small window of preparation before the Trials, Mrs. G made sure I still had her support. She organized fundraisers, updated the community on my health, and created the Blewitt Backers--friends and family wishing to share their encouragement as I trained for my dream. It was so comforting to look in the crowd at the Trials and within the sea of people and to see her among a group of my closest supporters wearing a t-shirt with a picture of me and "Blewitt Backer" in large letters across the back.

Coaches should motivate a sense of confidence in their athletes. Denise Gorski goes way beyond that effort. Her hard work sets an example for the whiny string bean girl who didn't want to lift weights (uh, that was once me). Her dedication commanded the attention of the coaching staff to inspire their athletes. Finally, Mrs. G's integrity was unmistakable when she removed a photo of her successful 4x800m girls relay team posing with Marion Jones. After the athlete's drug scandal surfaced, Mrs. G did not want to associate that tainted career with Boardman athletes. I am thrilled Allison has chosen Denise Gorski as an inspiration for her book, *Meathead: Unraveling the Athletic Brain.* Denise understood how the mind of the athlete must be nurtured, inspired and pushed to excellence. Her workouts were tough, she would never bend the rules, but in the end we all grew up stronger with her leadership. Our minds were as strong as our bodies in competition. We knew going into any meet that if we gave our best effort, Mrs. G would beam with pride and celebrate our victories. We didn't just win or lose and event. Mrs. G taught us to win or learn from the experience.

Adriane Blewitt-Wilson
13-time NCAA All-American in shot, discus, hammer, and weight throw
USA Track and Field Olympic Trials '04, '08, and '12
Four-time world champion of Scottish Highland Games

Prologue

As a scientist, I am the first to admit that the broadcasting of research to the general public can be convoluted. Research articles are littered with jargon and complex graphs that are, in turn, often misinterpreted by the media. I wrote this book with the intent of making the science relatable to you—the athlete, gym rat, and sports nut—by bridging the world of biomedical science with the athletic pursuits of myself and those of notable athletes.

I've been an athlete my entire life. I grew up in a community deep in the "Rustbelt" known for its world-class boxers, professional football players, and coaches: Youngstown, Ohio. As an infant, my dad would brag about my strength and balance because I was apparently able to hold a plank position in his hands. At the age of three, I began to participate in team and individual sports that greatly varied in their mental and physical attributes. I started with gymnastics and dance and then moved on to softball, basketball, skateboarding, and track and field. Over time, I became known as a daredevil. By my teenage years, I finally decided to narrow my focus on athletic pursuits that required infallible mental fortitude and presented the greatest risk of injury: gymnastics, pole vaulting, and hurdles.

After four years of competing in track and field in college, I decided that it was time to give back to the athletic communities that I had participated in by coaching for high school and collegiate teams. During this time, I also began doctoral work in a neuroscience laboratory specializing in circadian rhythms: persnickety physiological phenomena that cue you when to sleep, eat, exercise, and when to perform at your best. In particular, I study rhythms of sleep and activity in animal models and how these rhythms impact medical and psychiatric conditions such as stroke and drug addiction. Beyond the laboratory and classroom, I am still a competitive (and sponsored) athlete in the "Sport of Fitness": Crossfit®. Last year, I had the opportunity to qualify and compete in the Reebok Crossfit Games alongside a wonderful group of teammates who have also spent their lives pushing themselves to new mental and physical limits.

Moreover, the content of this book is driven by questions that I have had, but have rarely bothered to seek answers to across two decades of intensive training. The writing of this book is also driven by my passion for the scientific method, the perplexities of the brain, and the ability of the human body and mind to endure pain, perform well under pressure, and complete a series of complicated movements that sometimes require little thought. There are five chapters of this book that delve into the neuroscience of exercise, an emerging field of study, supplemented with autobiographical and biographical accounts from notable athletes.

By neuroscience, I mean that the scientific literature will strictly focus on the brain and how it responds and adapts to recreational exercise or regimens of intensive training. In the first chapter, I will explore the wide array of physiological events that occur in brain structures, brain circuits, and individual brain cells with general exercise. I will then document what makes the brain of a competitive athlete unique from that of a weekend warrior. From there, I will further dissect the brain of an athlete, highlighting the differences between an individual competitor versus team player. I will also provide a discussion on "muscle memory," the pre-workout adrenaline rush, and performance under pressure. From here, I will discuss my own research areas of circadian rhythms, sleep, and mental health. Lastly, I will discuss drug doping and how it affects the brain. I will talk about the widespread use (and abuse) of stimulants, hormones, and even how genes can tap into athletic performances.

I have catered my discussions to the fitness aficionado--someone who recreationally works out for personal reasons whether it's for vanity, general health and well-being, or the glorious workout "pump"; the competitive athlete—someone who *takes pleasure* in pushing themselves to new physiological and mental limits, goes above and beyond what their body can handle, and does this type of regimen day in, day out, rain or shine; and the sports nut—someone who has profound statistical and biographical knowledge of many sports, one sport, or a particular team and has no issue using their disposable income in support of their favorite sports team or city. As the reader, I hope that you no longer undervalue the brain when it comes to exercise and gain an appreciation for how easily the brain can adapt and positively respond to exercise. For athletes, I hope that you can view this as a handbook for gaining a competitive edge. The science of fitness and exercise has been around for decades, but the public has only become aware of this science within the past few years.

Introduction

By reading this book, I hope that you embrace the benefits of exercise for the brain and mental health, which are often overlooked by the "war on obesity" campaign. Beyond the benefits of exercise for mental health, I hope that you will be fascinated by how plastic the brain is. You will recognize that the brain is able to respond to new environments and stress remarkably well and that there can be near permanent changes in how the brain re-wires itself and communicates with the rest of the body just with practice. These adaptive changes do not just happen as a budding adolescent (although it helps), but can persist across adulthood! For those of you who feel as though you are destined to live the life of a couch potato who can only watch others bask in athletic glory, I hope that you will realize that your brain and health can be quickly remedied.

In writing this book, there were three demographics that I sought to target: the fitness aficionado--someone who recreationally works out for personal reasons whether it's for vanity, general health and well-being, or the glorious workout "pump;" the competitive athlete—someone who *takes pleasure* in pushing themselves to new physiological and mental limits, goes above and beyond what their body can handle, and does this type of regimen day in, day out, rain or shine; and the sports nut—someone who has profound statistical and biographical knowledge of many sports, one sport, or a particular team and has no issue about using their disposable income in support of their favorite sports organization or team. I hope that those of you who fall into one of these three demographics found knowledge and amusement in most, if not all of the chapters.

For the fitness aficionado, you should feel empowered by the widespread changes in the brain that come with daily physical activity, ranging from the improvement of mood, memory, learning, alertness, coordination, and reaction time. These are documented in chapter 1 (Barbells and Brains). I also expect many of you to start paying more attention to how much you sleep, when you sleep, and your quality of sleep as documented in chapter 4 (Eat, Train, Sleep). For the sports nuts, I hope that you appreciate the discussions on the underlying biology of memorable athletes, sporting events, and play-by-plays. In chapter 2 (The Athlete Brain), I hope that you find enjoyment in the recruitment of loyal and knowledgeable fans much like yourself to study where intense emotions experienced during a high-stakes game are regulated in the brain.

For competitive athletes, I hope that you view this as a handbook for gaining a competitive edge as you continue your athletic endeavors. You have spent years training your brain and body to react and adapt to physiological and mental stress. You deserve a boost. For young athletes, I hope that you find motivation from knowing the importance of practice for the brain's sake and how such re-wiring of the brain with practice can revolutionize athletic performance.

For high school and college athletes, you can certainly use this book as empirical support to refute being called a derogatory term like "dumb jock" or "meathead" by your non-athletic peers. These issues are discussed in chapters 2 and 3. Much like the fitness aficionado, I hope that you begin to pay attention to your sleep schedule, especially during game day travel, and realize how much and when you sleep can dramatically affect athletic and scholastic performances. I hope that you find solace in the discussion of pre-competition butterflies and realize that nervousness is perfectly natural. Also please don't overthink during a competition and rely on muscle memory that you have fine-tuned with practice instead; Your brain knows what it has to do from the 10,000 hours of practice. Let it do it.

Finally, I hope that athletes can take advantage of next generation performance enhancers provided by science in chapter 5 (Central Doping). Of course this requires careful research as to what you are putting into your body and using in moderation, but there is nothing wrong with using science to gain a competitive edge if it is within the guidelines. To conclude, I have tremendously enjoyed the research and writing of this book. It's not every day that I can relate my years of blood, sweat, and tears on the track and in the gym to scientific concepts or have the opportunity to express them to the greater athletic community rather than my own specialized scientific community.

Chapter 1

Barbells and Brains

I do not want to give you a watered-down, Cliff's Notes version of an exercise physiology textbook. This would be a huge disservice to researchers who have spent years laboring over massive amounts of data collected from millions of people. Instead, I want to broadly discuss the next generation of research in health and fitness. When we read about health and fitness in the newspaper, hear about it on early-morning news shows, and subliminally "see" advertisements for improving our health and fitness, what biological tissues are often the central focus? Muscles, fat, and the heart.

Yes, it is absolutely true that a toned body, calorie-restrictive diets, and a healthy heart help us live longer and look and feel sexier. But at the end of the day, the food we eat and the water we drink are primarily needed to fuel, hydrate, and keep one organ working twenty-four hours a day: the brain. The brain is without a doubt the most neglected organ of general health and fitness. When someone who loves to work their "bi's and tri's" is called a *"meathead,"* it is often referring to their physical appearance.

Although, calling someone a meathead can also be used to insult someone's intelligence (or lack of brain function).

Luckily, athletes, trainers, coaches, and sports nuts—people who closely follow a team's or athlete's statistics and performances across a season, win or lose, and understand the physically and mentally demanding nature of the sport—have different viewpoints. As a lifelong athlete and coach to high school and collegiate athletes, the terms "*mental block*," and "*brain drain*" are often used to explain some error or mistake made in play. To this end, the intent of this chapter and the general theme of this book are to convince you that the brain is equally as important--if not more--than the muscles and heart when it comes to improving our health—physically and mentally—with exercise.

Frog Goo and the Early Days of Brain Science

Exercise is a physically demanding process. Anyone who has fallen off or has tried to get back on the "fitness train" understands. The first few minutes of exercise are often the most difficult. During this time, our bodies go into a brief state of oxygen debt. Our hearts have to pump harder, our lungs have to expand and contract quicker, and blood has to travel faster to keep up with the increased gas exchange of oxygen and carbon dioxide going on in our muscles, organs, and cells. In order for exercise to continue, there needs to be a readily available and constant supply of sugar (i.e. glucose), fat, and production of chemical energy (i.e. ATP, adenosine triphosphate) in the mitochondria or the "powerhouse" of a biological cell.

This is the central dogma of exercise physiology. In looking through the notes and textbook of the exercise physiology course that I took in college under the late Dr. Leon Goldstein—a pioneer in the study of water balance in cells—the brain was discussed in one paragraph of one chapter toward the end of the textbook. I certainly do not want to undermine the vast knowledge of and appreciation for exercise physiology that I gained from taking Professor Goldstein's course. However, I do hope to make up for what is lacking in most exercise physiology textbooks by the end of this chapter.

The brain is a "chatty Kathy." It is constantly in communication with billions of nerve cells termed neurons at a single instance in time. When we are awake and active, there can be a lot of interference. We know this from the study of brain communication in the laboratory; most physiologists study brain activity and cross-talk between neurons by affixing electrodes around the scalp of a subject: mouse or human.

During wake and especially during periods of high activity, brain waves emitted from these electrodes are very fast and sort of look like a book editor furiously hand-proofing a book. During sleep, communication wanes and neurons take turns going "offline." At this time, brain waves are very slow and resemble a Kodak moment taken at the foot of the Rocky Mountains. This is one very basic way in which the brain when we exercise is different from when we sleep or "vegged out" while driving, watching TV, and doing something habitual.

The physical properties of neuron cross-talk are electrical and chemical in nature. We know this from studies undertaken in frogs and isolated organs starting in the eighteenth century. In the 1700s, an Italian scientist named Luigi Galvani was famous for making leg muscles removed from a frog twitch by means of an electrical current. Soon after, it was discovered that the muscles would not twitch in the absence of electricity or if the nerves around the muscle were removed. Because of Galvani's experiment, people were taught for a hundred years that nerves and organs communicated by means of an electrical current.

However, in the early 1900s, a German researcher—Otto Loewi—had a radical idea and re-wrote medical textbooks in the process. He surmised that communication between neurons was due to electrical currents coupled with chemical signals released by the neurons. The experiment to test this hypothesis came to him in a dream. In his famous experiment, he isolated frog hearts and kept them alive by incubating them in a solution within the range of normal physiological conditions. One of the hearts contained the frog's vagus nerve which when stimulated, slows down heart rate. The vagus nerve serves the same purpose in humans too. The other heart did not have a vagus nerve. First, Loewi stimulated the heart with the vagus nerve via an electrical current. This caused the heart rate to slow. He then applied some mystery fluid taken from this slowed-down heart to the heart without the vagus nerve. When this mystery fluid was applied to the heart lacking a vagus nerve, its heart rate also slowed! Viola!

Communication between neurons must be electrical and chemical in nature. This is the current dogma of most neuron communication. As far as the contents of this mystery fluid, it was later identified to be acetylcholine, which I will talk about later in regards to its importance for exercise.

Neurochemical Eruption

If you do not already know that the brain communicates by a vast assortment of chemicals, just watch late-night TV and you are guaranteed to see numerous commercials for prescription drugs that claim to help you sleep better or feel less anxious and depressed. These compounds—known as neurotransmitters and neuropeptides—are produced in each neuron and released when a certain electrical threshold is achieved.

They are released in normal quantities when life progresses smoothly, but are released like water from a dam--in a good way--when we exercise. There are dozens of neurotransmitters and neuropeptides. The major difference between a neurotransmitter versus neuropeptide is size with neuropeptides being larger. Both of these neurocompounds act to excite or inhibit the brain in a certain manner. Instead of focusing on the biochemistry of all neurocompounds, I want to focus on four—dopamine, serotonin, norepinephrine, and acetylcholine—that are widely studied in the fields of exercise physiology and athletic performance.

Dopamine is my go-to neurochemical of scientific study and personal activation. This is evident if you glance at my left forearm.

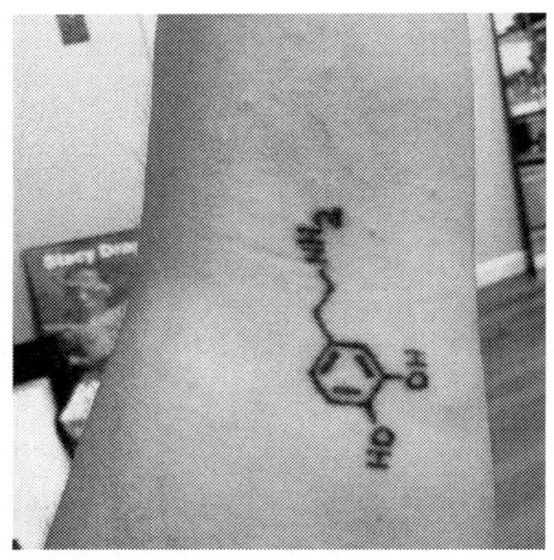

Embrace your inner dopamine: The neurochemical responsible for the rewarding, pleasurable effects of exercise. Photo by Allison Brager.

Why did I get this tattoo? After spending five years pursuing graduate work that often required me to stay up for thirty-six hours at once to collect samples every twenty minutes or go into work at 3:00 a.m. to inject alcohol or cocaine, I needed some type of tangible non-diploma to commemorate the experience of graduate school. While contemplating what tattoo I should get, I knew that I wanted a neurotransmitter that most accurately fit my personality and lifestyle. In the end, it was not a tough decision because dopamine is released from neurons in large quantities and high frequencies whenever we engage in any type of rewarding or motivational behavior: exercise, socializing, sex, eating, and doing drugs, to name a few. For me, I get my daily dose of dopamine by putting my body through mentally and physically challenging workouts, talking shop or joking around with colleagues and roommates, or eating lots of dark chocolate. In the fields of neuroscience and psychology, we refer to this as hyperhedonia. But for now, let's focus on the impact of dopamine on exercise, and not athletic performance. I'm reserving that discussion for the next chapter, which is titled "The Athlete Brain."

Dopamine is produced from the amino acid tyrosine. It is released from neurons in specific brain regions. From the careful study of rodents in the laboratory, we know that dopamine is recruited almost immediately during exercise[1,2]. Motivating hamsters and mice to exercise is extremely easy. If you put an unlocked running wheel into their cages, they will voluntarily run for an average of eight hours a day, during which time they will cover many miles. But hamsters and mice are not "weekend warriors." They will voluntarily run several miles day in and day out to the point that the researcher can accurately predict when they will run and when they will not.

In graduate school, I worked in a laboratory that studied wheel running in hamsters and mice. We not only had hundreds of hamsters and mice that would wake up and run on their wheels within five minutes of the lights going off on a daily basis--hamsters and mice are nocturnal--but some of our animals would run as much as ten miles a day!

It is also possible to measure a hamster's or a mouse's ungodly craving for exercise by putting them on a treadmill. Yes, a treadmill with divided cubicles to see how far and fast they will run until exhaustion. Both experimental approaches have been used in the fields of neuroscience to determine the time and amount of dopamine release from the brain. Measuring the release of dopamine or any neurochemical is accomplished by placing a diffusible probe into the brain region of interest.

With this probe, a researcher can inject drugs into the probe or sample biological substances that travel out the other end of the probe once a chemical equilibrium is reached. These types of experiments kept me up for thirty-six hours at a time in graduate school. Only I was looking at how drunk mice would get after giving them the equivalent of a half a bottle of cider (low dose), wine (moderate dose), or liquor (high dose) and measuring the amount of alcohol reaching their blood and brain. For researchers who have used these techniques to study the release of dopamine with exercise, the general consensus is that dopamine is drastically released at the beginning of exercise and maintained throughout[1,2].

Where is this dopamine produced and released? Researchers have spent decades determining which areas of the brain regulate the seeking and craving of all rewards: exercise, drugs, sex, and food. After years of careful probing, there is a general consensus that the mesocorticolimbic and mesopontine circuits of the brain and their release of dopamine drive the seeking and craving of rewarding stimuli.

Let's break down these circuits. Fortunately, neuroscience taxonomy is straightforward. Someone who knows the basic anatomy of the brain can figure out that brain areas controlling reward lie in the middle part of the brain stem (meso-), the outer, wrinkly layer termed cortex of the brain (cortico-), the middle of the brain (limbic), and a region of the brainstem known as the pons (pontine). The specific structures of these mesocorticolimbic and mesopontine circuits include the ventral tegmental area, the frontal association cortex, the amygdala, and the pedunculopontine tegmentum, respectively. These are a few of many structures regulating reward I might add. In brief, these brain areas work together to control the extent to which a reward is sought, craved, and satisfying.

Recently, we have discovered that the the reward circuitry does not discriminate between exercise and hard drugs like cocaine and methamphetamine[3,4]. It is possible for exercise to be just as addicting as abused substances. It is also possible for exercise to serve as a rewarding substitute for abused substances. This is a relatively new area of biomedical research. Several studies in rodents and populations of drug users have shown this phenomenon of exercise-drug substitution--also known as hedonic substitution.

In fact, I worked on a study of such sorts with a lab mate in graduate school. Hamsters love to run, but they also love their booze. Our calculations reveal that a hamster can <u>voluntarily</u> drink fifty times more alcohol than the average human male on a <u>daily</u> basis. This is largely because the types of hamsters that we worked with--the Syrian hamster--eat fermented cacti fruit in their natural environment: the desert. Thus, natural selection has allowed for the Syrian hamster to be a functional alcoholic.

With all this information at hand, we found that wheel running was a satisfying substitute for alcohol in hamsters a few months in age as well as in very old hamsters. In this study, the hamsters were presented with two scenarios—access to alcohol and a working running wheel or access to alcohol and a locked running wheel. When the wheel was unlocked, the hamsters of all ages drank less alcohol and ran more. But when the running wheel was locked, the hamsters drank more alcohol. These results are not a coincidence because other labs have reported similar results in mice[5]. We also know that these results can be extrapolated to humans. Exercise is a recommended and prescription-free approach to curb drug addiction in recovering addicts and can benefit individuals with a family history of drug abuses ranging from alcohol to cocaine[3,6].

From personal experience, I would consider myself an exercise addict. Even when I took a break from athletic competition post-college, largely because Crossfit did not yet exist and I was not yet considered a Master's competitor under the guidelines of USA Track and Field, I still needed to exercise daily and intensely. If I had not yet exercised past seven in the evening, I would become irrationally irritable.

In fact, I wish that I could bottle the euphoria that I experience after a grueling and painful workout or a session of heavy squatting. While my addiction to exercise is physiological, there are also plenty of people psychologically addicted to exercise via body dysmorphic disorder. As the name implies, these individuals are obsessed with looking a certain way or being a certain weight. It is particularly common among male athletes who strive to be of a particular muscular build, but also among females.

The addictive nature of exercise is not only driven by dopamine. The neurochemical serotonin is also influential. Serotonin is derived from the amino acid tryptophan; the amino acid linked to Thanksgiving turkey stupors. In athletic communities, serotonin is well known for being (loosely) responsible for the "runner's high." I say serotonin is loosely responsible for the "runner's high" because behavioral neuroscientists know that dopamine and a variety of other neural signals are also impactful. Regardless, serotonin is certainly recruited during exercise.

In fact, the impact of serotonin on wheel running and biological clocks in rodents has been a focus of my graduate mentors' laboratories for three decades. Drs. J. David Glass, Rebecca Prosser, and their students have identified the cascades of serotonin communication that occur in individual neurons, specific brain areas, and the space between nerve cells during exercise and what impact exercise and serotonin have on biological clocks. Aside from it being relatively easy to force a rodent to run during the middle of the day when they are likely to be sleeping, samplings of serotonin release over time or blocking serotonin communication with pharmaceuticals reveal a strong correlation with the extent and time of wheel running.

Studies that have used diffusible probes targeting specific brain regions reveal that serotonin communication increases when hamsters run during the middle of the day. Wheel running in itself increases serotonin communication in the “master” biological clock located deep in the brain in a region of the hypothalamus known as the suprachiasmatic nucleus (SCN) as well as in the major production center of serotonin—the raphe nucleus of the brainstem[7]. In addition to increasing serotonin, allowing rodents to run on wheels for a few hours during the day causes their normal routine of nighttime wheel running and exploring to occur sooner[8]. Studies that have used pharmaceuticals show that blocking or inducing serotonin communication can reduce or increase, respectively, the time difference in nighttime activity as well.

But what effect does serotonin have on exercise habits in humans? This question has not been answered directly, but we do know that antidepressants, which act to increase serotonin communication by blocking the retrieval of serotonin released from neurons, can impact levels of daily exercise and activity; most antidepressants are wake-promoting drugs since most of them increase serotonin communication. I do not want to belabor upon this, but the use of antidepressants to treat anxiety and depression is a double-edged sword.

One of the common points of the debate in the sleep research community is weighing the pros and cons of antidepressant use for the treatment of mild depression. This is largely because insufficient sleep worsens symptoms of anxiety and depression. Because most antidepressants are wake-promoting, antidepressants, in turn, can exacerbate rather than treat symptoms of depression and anxiety. Despite this conundrum, antidepressants can increase fitness albeit indirectly. Antidepressants do not proactively improve endurance, muscle strength, and agility in the absence of a fitness regimen. However, they do improve a depressed or anxious person's quality of life, which can increase the likelihood that one will exercise.

In the realm of athletics, antidepressants are used for a different reason; antidepressant abuse is common among collegiate athletes in order to deal with the pressure to perform and retain an athletic scholarship or to deal with the mental aftermath of an injury or overtraining[9]. Also, a significant proportion of retired professional athletes take antidepressants to deal with the depression of retirement. The risk of depression is even worse among professional athletes that have career-ending injuries.

The morbid fates of professional athletes make perfect sense. These individuals are at their peaks physiologically and socially; if you live in a sports-centric city like New York, Boston, and Atlanta, you likely understand this. But when these peaks in fitness and notoriety are stolen by a misfortunate injury or free agency with little preparation and guidance for adapting to the real world, the likelihood of a professional athlete developing depression is a critical concern.

Antidepressants are also prescribed as often as 25% of the time to adolescents who have suffered from a sports-related concussion[10]. Presently, the most talked about abuse of antidepressants in the sporting world concerns current and former players in the National Football League (NFL). Current players are given antidepressants to manage mental deficits post-concussions much like young kids. Retired players are commonly prescribed antidepressants to not only deal with retirement from their sport, but also to curtail high risks of suicide and major depression after years of inadvertent brain damage caused by persistently using one's helmet as a weapon on the field or getting brutally thrown onto the ground.

What happens to one's physical and mental well being in the absence of serotonin? The results from rodent studies are polarizing because daily energy levels can decrease while one's aggression can simultaneously increase. This is why antidepressants are commonly used in therapy for anger management. Depleting an animal's brain of serotonin can cause social havoc, especially in hamsters.

Hamsters love to run, they love their booze, and they also love to fight. One of the biggest concerns for hamster breeding is to ensure that the female does not kill the male after they mate. The female hamster's basic instinct to kill the male is not as cut throat as that of the female praying mantis, but she will kill her mate if he hangs around too long after they have finished only because she considers him a direct threat to her unborn young. Male hamsters are also extremely territorial towards other males. Researchers can set up hamster fighting rings by means of placing one hamster into another hamster's home cage. This "intruder" will immediately be pulverized if it doesn't fight back.

I relate all these peculiarities of hamsters to you because aggression in female and male hamsters is regulated by serotonin. Changing the release, binding, and retrieval of serotonin can have immediate and significant effects on aggression levels. But is this also true for humans? Unfortunately, there is not much research on the subject matter. Perhaps it is because performance-enhancing drug (PED) abuse is so highly stigmatized even in research communities. Part of me also wonders if depleting serotonin has been the trade secret to increasing aggression immediately prior to a mixed martial arts fight.

The last neurochemical of the euphoria trifecta relevant to exercise is norepinephrine. Similar to dopamine, it is produced from the amino acid tyrosine. In fact, the synthesis of dopamine precedes that of norepinephrine, which means that the amount of norepinephrine produced depends, in part, on the amount of dopamine produced. The major production center of norepinephrine is deep in the brainstem in an area known as the locus coeruleus. Much like serotonin, norepinephrine is wake-promoting. This is why pre-workout drinks that aim to rapidly boost energy levels are saturated with derivatives of norepinephrine. Many antidepressants target neurons that alter the release of norepinephrine in addition to targeting neurons that release serotonin. Beyond providing an energy boost, norepinephrine also increases blood flow. This is why athletes, including myself, will often take a few puffs from an inhaler before or during competition.

While dopamine, serotonin, and norepinephrine are popular in the world of exercise physiology and athletic training because of their euphoric properties, the neurochemical acetylcholine is known for being an enhancer of physical and mental performance. Acetylcholine is produced from the amino acid choline. It is the major neurochemical of muscle. It is what causes muscles to fire. It drives the strength of a muscle contraction—force—as well as the duration of contraction—endurance. It is clear to see why acetylcholine is a common target in the world of drug doping, is it not? In the brain, it is found in great abundance in the outer, wrinkly cortex as well as in areas of the brainstem that are difficult to pronounce: the pedunculopontine tegmentum and laterodorsal tegmentum.

Caffeine is one means to reap the benefits of acetylcholine for improved physical and mental performance. I do not want to elaborate too much here because I am reserving this discussion for the last chapter of the book—"Central Doping"—but caffeine is a fast-acting, effective, and popular performance enhancing drug. It increases the four most important skills of an athlete--alertness, focus, endurance, and power. Much like antidepressants, the use (and abuse) of acetylcholine for boosts in athletic performance can be a double-edged sword. Too much acetylcholine or inappropriately timed release of acetylcholine can disrupt sleep. As you will learn in the second to last chapter of this book—"Train, Eat, Sleep"—sleep and recovery is just as important as training.

The other importance of acetylcholine is that it is the neurochemical backbone of the sympathetic (SNS) and parasympathetic (PNS) nervous system. You may recall learning about the SNS and PNS in high school biology and even in the textbook, an athlete may had been used to exemplify the properties and functions of the SNS and PNS. The general idea is that the SNS is activated in times of stress like a competition. If you look into an athlete's eyes minutes before the game or race begins, their eyes are likely dilated. You may also notice some fast-paced breathing, involuntary hand shaking, and frequent swallowing (from dry mouth). This series of events prepares the body for action. In contrast to the SNS, the PNS is activated in times of rest and relaxation after the competition. The eyes are constricted, breathing is slow, and hunger is common. For both of these systems, acetylcholine is the go-to neurotransmitter that cues certain organs to do certain things under conditions of stress or rest.

To conclude, although there are many other neurochemicals that have *some* impact on exercise, these four neurochemicals—dopamine, serotonin, norepinephrine, and acetylcholine—are most widely studied by neuroscientists (of exercise) and manipulated for consumption in powder and liquid form by legitimate and underground PED companies. We will hear much more about each of these PEDs in "Central Doping."

Biological Juicing, Pre- and Post-Workout

This book would not be complete without a discussion of adrenaline. Instead of describing the adrenaline rush because I guarantee 95 percent of you are very familiar with it, I want to talk about its purpose. Adrenaline is needed to mobilize fuel for the muscles, brain, and organs quickly and broadly. It stimulates rapid breakdown of the three basic fuel sources—glucose, fat, and protein—which is then shuttled to the areas of the body that need it most.

All of these physiological phenomena occur well before it is consciously recognized. Adrenaline is not produced in the brain, but the adrenal glands. The trigger for adrenaline release originates from the brain. Note: this will not be the first time in this book that you hear about this physiological cascade of events known as the hypothalamic-pituitary-adrenal (HPA) axis. Nearly every chapter has some focus on adrenaline and its importance to athletic performance, recovery, and gaining a competitive edge.

The first brain area leading to the adrenaline rush is a region of the hypothalamus known as the paraventricular nucleus (PVN). The PVN releases corticotropin-releasing hormone (CRH). This signals then trickles down to the pituitary gland, which releases adrenocorticotropic hormone or ACTH for short. From here, the signal travels outside of the brain and the adrenals and appropriate organs are informed via the release of adrenaline and the activation of the sympathetic nervous system. Behaviorally, adrenaline greatly benefits an athlete. It increases focus, allowing an athlete to tune out the roar of the crowd. It increases power and stamina and if necessary, aggression. It can also be annoying if there is a lag in athletic competition following an adrenaline surge. I will discuss more of this in the “Athlete Brain” chapters.

If adrenaline is a pre-workout juicer, what stimulates tissue recovery and repair post-competition? One post-workout or competition "juicer" that keeps the mind and body fit are cytokines. They are a unique group of proteins or proteins with carbohydrate attachments—glycoproteins—that increase or reduce inflammation. Inflammation is a general term to describe a series of complex biological processes that occur in response to some environmental or internal stress. It does not just describe swelling around the site of an injury or body part that is overused. This is a topic for another day, however. An increase in inflammation can ward off illness and disease whereas a decrease in inflammation promotes healing. There are several cytokines that fit within one of these two categories with other "go"-"no go" biological molecules that instruct how much of these pro- and anti-inflammatories are to be released.

The most commonly studied pro-inflammatory cytokines are the interleukins, particularly IL-6 and TNF-alpha. These interleukins are released by immune cells and act on several tissues of the body and brain. In the brain, most attention has been paid to the hypothalamus. IL-6 acts in the hypothalamus to cause a change in core body temperature, most often in response to sickness; a change in IL-6 expression in the hypothalamus is what causes you to feel clammy or overly hot when sick. IL-6 is also recruited during exercise especially with high intensity, strenuous exercise where there are rapid changes in body temperature, heart rate, and blood flow.

My colleagues and I have studied IL-6 in the lab. I will discuss this study in more detail in the chapter entitled, "Train, Eat; and Sleep" because of the relevance to jet lag and its impact on sleep and athletic performance. In brief, we found that mice that were routinely exposed to monthly trips around the world—by shifting their light-dark cycle eastward by six hours every week for one month—released more IL-6 from immune cells in the blood in response to a toxin known as lipopolysaccharide[10]. Further, this increase in IL-6 in response to the toxin was only found in mice that went on this trip around the world and not in a separate group of mice that were only deprived of sleep (but not exposed to jet lag) in amounts comparable to the world travelling mice.

In the world of athletics, routine travel across time zones during the competition season is a likely culprit as to why many athletes come down with some bug during the season even if the players are sleeping enough. In fact, I predicted that increased travel across time zones was one of the many reasons why players in the National Basketball Association (NBA) were coming down with sickness or injury during the truncated 2012 season, which resulted in fewer games but more frequent travel. This prediction came to fruition in our nation's major newspapers--the *Washington Post* and *LA Times*--when medical experts also stated that frequent travel was the likely reason behind a season of sickness, injuries, and poor refereeing.

Another pro-inflammatory that acts in the hypothalamus is TNF-alpha. Instead of changing core body temperature, TNF-alpha activates the HPA axis especially in times of shock. This is exactly what occurs after an athlete experiences a muscle injury. However, pro-inflammatory cytokines like TNF-alpha are not the only biological factors important for tissue repair and recovery. Anti-inflammatory cytokines also have a role.

Anti-inflammatory cytokines such as interleukin-10 (IL-10) are recruited not only during exercise, but initiate constant rebuilding and repair post-exercise. IL-10 serves as the body's natural anti-inflammatory, but when looming injury and overtraining do arise, the body needs something more like Advil or Motrin--a non-steroidal anti-inflammatory (NSAID)—to accelerate the process of tissue recovery and repair. NSAIDs damage the liver when taken over extended periods of time, but they do wonders to eliminate the feeling of being hit by a train during competition like I felt at times during the 2013 Reebok Crossfit Games.

Overall, studies in the world of exercise physiology have found that the body creates a balance between the releases of pro- and anti-inflammatory cytokines that then act on the brain during exercise[11]. As the intensity of exercise increases, so does the release of these pro- and anti-inflammatory cytokines. Part of me wonders if a fine-tuned balance of pro- and anti-inflammatory cytokine release is the biological reason that underlies why I have gotten sick less frequently since starting Crossfit. This anecdotal evidence has been relayed by other Crossfitters. Whether or not Crossfit re-wires and optimizes the balance of pro- and anti-inflammatory cytokine release remains to be determined, but I will say that the balance goes awry in the face of overtraining or jet lag.

To focus on jet lag, a disruption in the balanced release of pro- and anti-inflammatory cytokines is one of the main findings of a study that preceded the IL-6 study that I worked on[12]. Much like the IL-6 study where we sent the mice on a month-long trip around the world, a variety of different pro- and anti-inflammatory cytokines were measured in response to a month-long trip around the world coupled with the presence of a toxin. Jet lag caused simultaneous increases in the release of pro-inflammatory cytokines such as IL-6 and decreases in the release of anti-inflammatory cytokines such as IL-10. I will talk more about the dangers of jet lag for athletic performance and recovery, but the bottom line is to train smart, travel smart, and sleep smart.

The other post-workout "juicers" that I want to discuss are neural growth factors. As their names imply, growth factors contribute to repair and maintenance of the central nervous system. The two most common growth factors studied in the realm of exercise physiology are nerve growth factor (NGF) and brain-derived neurotrophic factor (BDNF). Growth factors help maintain normal operations of the neurons that can range from helping the neuron perform a specific task to guiding the neuron's pre-determined death. These growth factors are also heavily recruited when injury to a nerve or tissue such as muscle arises.

BDNF has been a huge focus in exercise physiology. Numerous studies have found that exercise promotes the production of BDNF arising from the endoplasmic reticulum—a major avenue in protein synthesis--in the neuron. This increase in BDNF production, in turn, fosters the growth of neurons by allowing them to communicate with more neurons, and can also protect the neurons from inadvertent or (un)pre-determined cell death[13]. This is one reason why BDNF is an attractive candidate for the treatment of dementia. It is also another reason why exercise is a viable treatment option for reducing risks of dementias. Exercise is highly encouraged for individuals with a strong family history like me.

Beyond dementia, BDNF is studied by scientists interested in environmental enrichment, which is basically any physical, mental, or social activity that stimulates the brain. Most of these studies have been conducted in rodents. These studies show that providing a mouse or hamster with the opportunity to run on a wheel increases neuron communication via BDNF. The same benefit applies for rodents that are presented with novel toys or playmates (even for male hamsters).

In a world saturated with social media, a craving for attention, and other forms of environmental enrichment, I would be curious to see how our obsessions with texting, tweeting, and posting has transformed the production of BDNF in the brain. To date, most of the psychology and biology-based research on social media has focused on negative outcomes such a jealousy and other "green-eyed monsters." I would also be curious to see how BDNF production in a lifelong athlete or fitness aficionado compares with the general, less fit population. At any rate, the results are likely to be positive for the fit population given what we know in other mammals.

From Molecules to Behavior: Mood and Memory

By now, you may think that I have become too obsessed with the details of the biological systems of the brain that benefit from exercise. A majority of the studies that I have cited so far come from simpler mammals that are elite athletes on their own terms. These simpler mammals have allowed us to understand how the biology of the brain of a more complex mammal like us changes and adapts to exercise. Technology is developing to address these unsolved mysteries of the human brain, but it will be years before many of these questions are answered. Now there's no need to worry. Exercise does benefit our mental lives. I will elaborate to an extent here, but the next chapter of this book—"The Athlete Brain"—will reveal all.

If we think about the general qualities of an athlete and what separates a mediocre versus great athlete, most of us would likely agree on the following buzzwords or mental attributes: reaction time, coordination, perseverance, decision making, risk taking, and problem solving. All of these qualities derive from activities of the brain and neurons. Reaction time and coordination require little conscious thought and improve with experience and practice as you will see in "The Athlete Brain."

The neurological foundations of reaction time and coordination arise from unique groups of nerves—alpha motor neurons and Golgi tendon organs—that innervate our tendons and muscles and communicate with the spinal cord and brain to determine how much we should move and where we are in time and space. These neurological foundations are also the basis of "muscle memory" which I will discuss in the chapter titled, "The Athlete Brain Continued."

The latter three attributes of an athlete—decision making, risk taking, and problem solving—are more often scrutinized by sports commentators than general sports fans. While you may think that these three attributes require significant conscious attention and mental processing, most seasoned athletes would disagree. Yes, football players and other team sport athletes spend several hours a week reviewing film from a previous game or for a future opponent, but decision making, risk taking, and problem solving during the midst of competition is largely instinctual and subconscious.

Winning is dictated by experience and practice, which is why head coaches are more inclined to keep their injured first-string quarterback in the game rather than bring in their healthy second- or third-string quarterback. Moving away from athletes, there is no denial that exercise for the weekend warrior or daily enthusiast improves mental well being so let's discuss this further.

Earlier in this chapter, I stated that exercise has evolved to be nature's antidepressant. This is because exercise has the same effect as antidepressant medication, behaviorally and physiologically: it improves our mood and increases serotonin communication. Data from thousands of questionnaires, interviews, and studies conducted in a clinic or laboratory reveal that a weekly regimen of exercise can decrease risks for depression and anxiety. It can even curtail substance abuse and delay the onset of dementia. These benefits of exercise for mental health are even stronger for individuals with a family history. However, instead of focusing on what types of exercise one should be doing, I want to focus on a biological circuit of the brain that is largely responsible for these decreases in depression, anxiety, and even dementia: the limbic system.

The limbic system lies in the more archaic areas of the brain: the brain develops inside out much like a blossoming flower. This blossoming metaphor also applies to changes in brain structure and function across millions of years of evolution in mammals, reptiles, birds, amphibians, and fish. "Smarter" animals are identified by the size of the brain's outer, wrinkly cortex. Hypothetically, the more cortex an animal has, particularly in the frontward areas of the brain termed the neocortex, the more complex the animal is intellectually, emotionally, and socially. While less intellectually adept animals such as reptiles and amphibians lack a sophisticated neocortex, they do have a limbic system.

The limbic system is comprised of several brain areas, but the two key structures are a bean-shaped structure known as the amygdala and a seahorse-shaped structure known as the hippocampus. These two brain areas work together to control emotion and memory.

Individuals who exercise regularly are, in most cases, thought to be more emotionally balanced. I am careful with my assertiveness because athletes are often stereotyped as being more emotionally volatile and aggressive. A "meathead complex" of such sorts may be confounded by steroid use and doping--a condition known as "roid rage"--or may also manifest from intense athletic training; training increases testosterone release in both men and women to allow the body to repair and recover in a timely manner. Disregarding athletes, exercise enthusiasts are postulated to have a better functioning limbic system with better "go" and "no go" controls for managing anxiety and despair. This has been shown in humans on several occasions.

Observing anxiety and despair in a human is fairly straightforward. We fill out questionnaires, undergo interviews with licensed professionals, and perform tasks in a laboratory wherein our performances are highly predictive of our emotional states. We can also be put in a brain scanner known as a functional magnetic resonance imager (fMRI) or receive an IV injection of radioactive dye and then undergo a brain scan—an experimental approach known as positron emission tomography (PET)—to determine how active our limbic systems really are, and as to how our limbic systems behave when we are asked to perform some task relevant to anxiety and despair.

Using experimental approaches such as these, researchers have found that a regimen of aerobic exercise or meditation-based exercises much like yoga or tai chi effectively calms individuals, behaviorally and physiologically[14]. Both exercise regimens were notably helpful for individuals with social anxiety disorder, which is a fear of behaving in a manner that could lead to public embarrassment or shame.

In brief, the researchers found that subjects who engaged in aerobic or meditation-based exercises were less reactive to negative stimuli generating embarrassment and shame. At the level of the brain, MRI scans revealed complementary changes in activity in frontal and parietal regions that regulate attention. Moreover, whereas anxious individuals were previously unable to consciously recognize or assuage their anxieties, a regimen of aerobic or mindfulness-based exercises transformed (re-wired) their brains in a manner that allowed these individuals to consciously recognize their anxieties and cope with imminent social threats under their own volition.

It is also important to study anxiety and despair in response to an exercise regimen in laboratory animals. The ability to identify adaptive changes in the brain and behavior with exercise in genetically-engineered mice can further help us identify the physiological benefits of exercise and even mine for a "fitness" or "athlete" gene. There are several ways to measure anxiety and despair in the lab.

During my postdoctoral fellowship, we measured anxiety and despair (although in the absence of an exercise regimen) by means of a Morris Water Maze. Mice hate water unlike their laboratory "sibling"—the rat. If you drop a mouse into a bucket of temperate water, it will often freak out and immediately try to escape. Luckily for mice in a Morris Water Maze, there is an elevated platform that the mice can crawl onto. So what does this have to do with levels of anxiety and despair? If a mouse is anxious or depressed, it usually takes them longer to find the platform; they will either haphazardly swim about or give up and tread water. You can also measure anxiety and despair in lab rodents through an elevated plus maze. A mentally healthy rodent will more readily explore many hallways and arms of the maze whereas an anxious or depressed rodent will cower in one corner.

Several studies in rodents have shown that providing a rodent with a running wheel and letting the rodent run under his own terms reduces anxiety and despair[15]; the animals are more willing to explore unfamiliar or open areas of a maze and show more mental fortitude when despair or defeat is measured. These observations are similar to that which you would observe in a seasoned athlete, are they not? Of greatest fascination, preventing a rodent from running on their wheel (by locking it) is even more detrimental to their mental health than not presenting the rodent with an opportunity to run at all!

In fact, rodents deprived of running in these studies have also been reported to be more aggressive towards their peers and human handlers; you may recall that my lab mates and I did a study in graduate school where we gave hamsters simultaneous access to a running wheel and alcohol. When we locked their running wheels, we also noticed that the hamsters became more aggressive towards us whenever we had to clean their cages. These findings—both empirical and observational—recapitulate how exercise can transform into a drug, at times.

Running studies in rodents are also beneficial for identifying genes and other molecular factors that control how anxious, defeated, or hyperactive a rodent is. A study of such sorts is referred to as quantitative trait loci (QTL) study in which a particular behavioral “phenotype” such as daily wheel running or physiological “phenotype” such as daily sleep amounts is measured in hundreds of mice with different genetic landscapes (genotypes). Because mice are inbred, meaning that brothers are often bred with their sisters, genetic diversity is low. Because genetic diversity is low, a phenotype such as daily wheel running or sleep amounts is less likely to vary within a particular strain of mice.

Thus, QTL studies are able to link some level of anxiety, defeat, or hyperactivity with some gene or cluster of genes on a particular chromosome. For example, a QTL study that examined daily rhythms of wheel running in three different inbred mouse strains found that the time at which the mice began their daily schedule of wheel running was due to a particular location on chromosome 2 (mice have 20 chromosomal pairs which is three less than us)[16].

The process of comparing frequency of exercise (phenotype) with genotype is also possible in humans. This process is referred to as a genome wide-association study (GWAS) with the DNA coming from spit. A GWAS was completed in 2009 in thousands of Dutch and American (Omaha, NS) subjects to identify genes associated with willingness to exercise. Questionnaires that documented the frequency and intensity at which the individuals exercised were documented. In general, 50% of the Dutch and 63% of the Americans regularly exercised. Analyses of DNA found roughly 435,000 genetic variations known as single nucleotide polymorphisms (SNPs) in the Dutch and 381,000 SNPs in Americans that corresponded with an exerciser versus non-exerciser[17]. Clearly there is more than one gene regulating fitness or athleticism. I will save more discussion for how genes can regulate how much we exercise and how successful we are at exercise for the last chapter of the book—"Central Doping"—where I will discuss the potential for "gene doping."

In addition to studying the benefits of exercise for emotional stability, a significant amount of research has also been done in the field of memory. I'm fairly confident that you have heard or read that exercise boosts memory. This has been a hot topic in the world of geriatrics because of the increasing rates of dementia—both Alzheimer's and general dementia—in first-world countries. There are several biological reasons as to why elderly individuals who regularly exercise more often have a better memory than their sedentary counterparts so let's talk about them. First off, I should say that our memory is not just measured by how many phone numbers that we know, movie dialogues that we can quote, or names that we can recall.

There are several types of memory—declarative, episodic, procedural, and emotional. Phone numbers, movie dialogue, and names would fall into the category of declarative memory. It's the memory of facts. Autobiographical and biographical accounts of life would fall under the second category of memory—episodic. The third category—procedural—will be most widely discussed in this book because procedural memory is what creates and maintains one's athleticism. It's the memory of learning how to ride a bike (and making it hard to forget how to), how to throw a baseball, or even do something as technically challenging as pole vaulting. The last category of memory—emotional—is a bit more complex. It allows us to remember particular facts and events based on some intensive emotional state. For example, many of us can accurately recall in vivid detail where we were, what we did, and how we felt on September 11th, 2001. Despite this complexity, all of these categories of memory are mastered and controlled by the limbic system and in particular, the hippocampus.

The hippocampus is the most commonly studied brain area of memory research in mammals. Changes in its size, architecture, and function are also extremely predictive of performance on a memory-related task; a smaller hippocampus corresponds with poorer performance on reciting words, tapping a sequence of numbers on a keyboard, and recalling events of a movie or story. The same is true for mice too. Mice with damaged hippocampi perform poorly on the Morris Water Maze trial-after-trial because they constantly forget where the platform is. The anatomical and physiological reasons behind these declines in performance on a memory-related task can be several-fold: 1) there aren't as many neurons that are able to communicate with each other; 2) the speed of the communication is slower; and 3) the structures or steel beams that support a neuron in space are weak, to name a few.

In the world of exercise physiology, the budding of new or regenerated neurons in the hippocampus has been a huge biological focus in rodents and humans. In rodents, the approach has been to count the number of new connections induced by an exercise regimen and how "biological juicers" like BDNF that increase in production with exercise fine-tune communication between neurons. In humans, the focus has largely been on output; how exercise enthusiasts perform on a declarative, procedural, episodic, or emotional memory tasks relative to non-exercisers as well as how an exercise regimen corresponds with changes in the size—surface area—and activity of the brain.

Aside from the hippocampus, a specific neurochemical system also deserves recognition in regards to how exercise can benefit memory and can even delay the onset of dementia. This neurochemical is acetylcholine. Earlier in the chapter under the section titled "Neurochemical Eruption," I talked about acetylcholine being a major driver of muscle contractions and alertness. Acetylcholine is also a neurochemical target in Alzheimer's research. One of the hallmarks of Alzheimer's is the breakdown of neurons releasing acetylcholine. The process begins with the destruction of the neurons' structural beams called microtubules. Because of this, the neurons releasing acetylcholine become tangled and die. In neuroscience communities, these neuron clusters are known as neurofibrillary tangles.

Because Alzheimer's is one of several dementias that fall along a spectrum of symptoms and severities, not every case of dementia is due to neurofibrillary tangles. In fact, these neurofibrillary tangles are a key biological identifier of Alzheimer's. Nevertheless, a truly accurate diagnosis of Alzheimer's does not come until after death; brain scans and other neurological examinations can be used to diagnose Alzheimer's to an extent, but the identification of neurofibrillary tangles and other forms of neurodegeneration does not come until the brain is dissected, dyed, and analyzed after a decade or two of struggle.

Drugs targeting acetylcholine centers are often prescribed in an attempt to delay the onset or forestall the progressive decline in neuron communication characteristic of Alzheimer's. But exercise is also recommended; exercise keeps the system, enzymes, and other biological factors responsible for the production and release of acetylcholine properly maintained and efficient. There are tens of studies ranging from extremely large—a few thousand—to small populations of study that both show that regular exercise can delay the onset or slow the decay of mental function in Alzheimer's as well as other forms of dementia. Exercise is particularly imperative and highly recommended for individuals like me with a strong family history of Alzheimer's. To this day, we don't have a cure for Alzheimer's, but the neural pathways that are most often recruited by exercise are thankfully the systems of interest to drug companies.

Much like Alzheimer's, exercise can also benefit individuals with a family history of Parkinson's. This neurological disorder arises from the breakdown of neurons releasing dopamine, particularly in a frontal area of the brain known as the substantia nigra. The substantia nigra is part of a motor feedback loop that directs and controls muscle movement. Several studies in rodents have shown that exercise regimens are "neuroprotective" against Parkinson's by keeping dopamine production centers, particularly in the substantia nigra, intact; much like how exercise keeps the acetylcholine system up and running in the case of Alzheimer's. I will talk more about the motor loop that falters with Parkinson's in "The Athlete Brain in Competition."

Unlike Alzheimer's, there is an effective treatment for Parkinson's in the form of a drug known as L-Dopa which increases dopamine communication. But not all treatments for Parkinson's are effective. Sadly enough, Muhammad Ali—one of the most physically coordinated athletes of the twentieth century—suffers from Parkinson's. Whether his condition manifests from a genetic mutation linked to Parkinson's or is from too many blows to the head is unknown. Further, while exercise may be sufficient to delay or forestall destruction of dopamine production centers in the case of Parkinson's, it has to be incredibly challenging to get individuals with a severe form of Parkinson's to exercise. Parkinson's is a movement disorder that arises from impairment in the brain—the master and controller of all movements big or small—and not the muscle. At any rate, as treatments for Parkinson's continue to be developed or improved, I imagine that the neurochemical targets examined will continue to be those activated by exercise.

Exercise for Traumatic Brain Injury

The topic of exercise and traumatic brain injury (TBI) raises eyebrows. In most cases, the conversation is one-sided in that there is little discussion on how leisurely exercise can improve daily life following TBI. Of course, most of the present discussion is on how extremely physical and violent exercise increases risks for TBI. But, it is important that millions of dollars continue to be donated by sports organizations and private donors in order to understand the timeline of risk for sports-related TBI in practice or play.

Let me clarify that the importance of sports medicine does not lie in scare mongering that encourages communities to ban youth sports like football or to make it "safer" by banning certain tackles. Indeed, the misuse of safety equipment as weapons can be more dangerous for players than tackling, and there is less awareness of this issue. That is my perspective.

In fact, the athletic department of my alma mater, Brown University, was one of the first universities to monitor an individual player's risks for concussion across the competition season during my time as a student-athlete. A next-generation sensor was implanted into each Brunonian's (derives from our mascot—the bear—and its name—Bruno) helmet and was interfaced to a computer. The sensor would track the frequency and intensity at which the player was hit in a single game and across multiple games. If the athlete had been involved in a dangerous tackle that potentially caused a "micro-concussion"—in which the brain is slightly rattled inside the skull—this hit along with previous and future hits of similar intensity would be logged. If enough micro-concussions occurred, the athletic trainer would be notified.

The identification and discussion of micro-concussions has become a buzzword in popular media. The idea is that one micro-concussion is not likely to lead to a full-blown concussion, but a series of them over a small window of time can certainly lead to a full-blown concussion. This is why there is so much concern over soccer players heading a ball traveling at high speeds or defensive linemen using their helmets as a weapon. However, what happens if TBI does arise from too many late helmet hits or pole vault crossbars to the head (as in my case) and a condition known as chronic traumatic encephalitis (CTE) develops? Can less violent physical exercise that does not involve using the head as a weapon (football), target (hockey), or extra limb (soccer), help delay the decline in brain function?

As I have discussed on several occasions throughout this chapter, exercise is a form of environmental enrichment, behaviorally, physiologically, and anatomically. It improves general well-being by reducing stress, curtailing despair, providing euphoria, and improving our memory, to name a few. Physiologically and anatomically, it promotes the growth of individual nerve cells, connections between nerve cells, and the size of whole brain areas. In biomedical literature relevant to TBI, exercise is characterized as "neuroprotective." What this means is that intermittent mild stress on the brain, such as a lack of oxygen (hypoxia) or blood flow (ischemia), causes adaptive changes in brain function, resulting in less damage to the brain when TBI does actually occur. Intermittent hypoxia is an attribute of exercise, especially intense exercise. Regardless of the type of exercise, the body spends the first few minutes of exercise in oxygen and blood debt, which is essentially a time for the heart, muscles, lungs, and powerhouses of the cell—the mitochondria—to acclimate.

In fact, I believe that oxygen and blood debt is the reason why the two most notorious workouts of Crossfit--"Fran" and "Grace"—are the most physiologically memorable (i.e. painful). The time window that it takes to complete "Fran" or "Grace"—which ranges from two to four minutes in a seasoned Crossfitter—also coincides with the time window of oxygen and blood debt

Measuring elite fitness: I was a research subject in a study at Kennesaw State University that aimed to measure aerobic and anaerobic performance in elite Crossfitters. Photo by Allison Brager.

While the elite Crossfitters in California were performing "Fran," real-time measures of heart rate and gas exchange were projected on a screen behind the athletes. Not only does one's heart rate and breathing reach maximal levels quickly into the workout, but it stays elevated until the workout is completed. Because the body can't get enough oxygen to support itself and the heart is manically pumping blood to the muscles, lactic acid accumulates, resulting in intense muscle throbbing and residual abdominal and lung pain for up to ten minutes after "Fran" or "Grace" is finished. Many individuals have puked. Although it has yet to be studied, I would wager that hypoxic and ischemic workouts like "Fran" and "Grace" could be neuroprotective.

As for the mechanisms through which hypoxia and ischemia relevant to exercise offer neuroprotection from TBI, this is widely inconclusive. Fortunately, the department that I worked in as a postdoctoral fellow has recruited faculty to investigate the mechanisms of neuroprotection in both animal models and clinical populations.

Personally, colleagues and I have studied neuroprotection from stroke in mice that lack one of the genes controlling biological clocks and sleep: *Bmal1*. Mice lacking *Bmal1* also do not exercise very much. They are extremely hypoactive, exploring much less and running on their wheels much less, than normal (wild-type) mice. They also have tons of skeletomuscular problems ranging from weak, contracting muscles to arthritis. By studying mice lacking *Bmal1,* we have found that these mutant mice have significantly more brain damage from a stroke compared with the normal mice even when exposed to mild ischemia beforehand[18].

If we extrapolate this research conducted in rodents to humans, there are numerous studies that identify exercise as a means to improve recovery and the quality of life following stroke, but little is known about how *exactly* exercise regimens are neuroprotective against strokes. However, I would predict that the mechanism involves one, if not more, of the neurochemical, physiological, and genetic factors of exercise that I discussed in this chapter.

To conclude this chapter, I hope that you have gained an understanding of the vast and adaptive changes in the brain that occur with exercise, regardless of the intensity. Exercise does not just affect a single brain center, but several. It improves our general psychological and physiological well being, decreasing risks for many psychiatric and neurological conditions. Exercise can alter physiology and nerve cell connections in a manner that protects the brain from further damage following injury. In the next few chapters, I will shift my focus to more specific demographics of exercise using the same principles and practices of exercise described here.

Chapter 2

The Athlete Brain

Ever wonder what goes through the mind of an elite quarterback when his team is down by 6 and there are three seconds left on the clock? Or someone who has spent most of their twenties on the couch, smokes, and becomes an endurance athlete in their forties? In this chapter, I will primarily explore the inner-workings of the brain of someone who recreationally works out and contrast their brain with that of a competitive athlete who spends a significant portion of their day pushing themselves to new physiological and mental limits. We also can't ignore the "sports junkie"—someone who may be a couch potato, but nevertheless, spends a large portion of their free time and disposable income displaying their allegiance to a professional sports team or city of sports. Lastly, I will explore what it takes to enter "the pain cave" and push through a grueling workout common of strength and conditioning programs.

The Mannings, Papale, and Rudy

There are many reasons why children, teenagers, and adults exercise. Most children have no choice. Parents enroll kids in individual sports like gymnastics and dance to teach them balance and coordination or team sports like football (American or European) to learn and fine-tune basic human social and instinctual skills. By one's teenage years, there begins to be a clear delineation in who works out and competes for pleasure or because of parental pressure. In the case of helicopter soccer moms and hockey dads, it has been shown that a child's level of enjoyment for a sport can be predicted by and is negatively correlated with the extent of parental pressure[1]. I've seen this occur on several occasions throughout my athletic career, even as a collegiate athlete.

During adulthood, the choice to exercise is often liberated from parental pressure and athletic competition and more often revolves around vanity and maintaining one's physical and mental health. This is definitely the ethos of my current "sport"--Crossfit--and likely explains its rising popularity amongst previously sedentary adults. Crossfit can cause a rapid and sustained re-distribution of one's fat-to-muscle ratio, improvement of cardiovascular health, and lifestyle choices all through performing "constantly varied movements at high intensity" and eating like a caveman; meats, veggies, nuts, some fruit, and little processed foods and dairy. While the shift in if and why someone chooses to recreationally exercise seems to be driven by personal health goals or societal standards, can one's intention to exercise more intensely for a different reason (i.e. competition) and success with athletics ever be "hard-wired," requiring little thought and motivation?

To address this, we need to look at the top layers of the cortex and deep into the brain in areas where basic motor skills are controlled and motivation stems. But to first put one's potential for athletic success into perspective, I'd like to introduce you to three professional athletes of America's favorite pastime whose rise to athletic success and fame came about from very different paths.

If your goal as a parent is to breed professional (American) football players, then I would recommend close inspection of the Manning family. Archie Manning is one of the more notorious quarterbacks of the 1970's who played for the New Orleans Saints, Houston Oilers, and Minnesota Vikings. He also begot two sons—Peyton and Eli—who would go on (and continue) to rack up numerous MVP awards and Super Bowl rings as quarterbacks. As I write this book, Peyton had a remarkable two years and Super Bowl appearance as quarterback for the Denver Broncos after his former team—the Indianapolis Colts—had a full-blown collapse the year prior when Peyton was on the sidelines recovering from neck surgery. Peyton's younger brother—Eli—has many accolades of his own having been a two-time Super Bowl champion and MVP. Sports pundits like Tony Kornheiser love to reminiscence about Eli's high school and early college years when Archie Manning would tell people "*wait to you see his [Peyton's] younger brother*" in reference to Eli's superb ability to throw the football and work the pocket.

I bring the Mannings' to your attention because they are a statistical anomaly. The likelihood that your *only* two sons, let alone three people from the same family, play in the NFL is exceptionally low. You can appreciate the Mannings' as being an exception to random probability if you consider the total number of collegiate players eligible for the NFL draft in a given year--roughly 65,000[2]--and the total number of players drafted--roughly 220, with each of the 32 teams in the NFL getting seven picks. Thus, independent of player potential, position, and college attended, a collegiate football player has about a 1% chance of suiting up in an NFL locker room. The likelihood of suiting up for a *single* position and the most coveted position of quarterback is even lower.

Although the Mannings' are a single case study and not part of some larger experiment aimed to identify the biological versus environmental factors contributing to the likelihood of becoming an elite quarterback, I would argue that their dominance had to be influenced by more than just day-to-day exposure to the sport as an infant, child, teenager, and adult, and may also be attributed to some unique "hard-wiring" in their brains. That being said, I am curious if anyone has ever thought to have the Manning clan undergo a brain scan to compare their brain anatomy, connections, and activity with other elite and rookie quarterbacks. That would sure generate high viewer ratings on ESPN.

There's also the story of Vince Papale whose life is portrayed in the movie *Invincible*. Vince was a star athlete in football, basketball, and track and field in high school, but only competed in track and field in college because his university, St. Joseph's University in Philadelphia, did not have a football team. After college, he returned to the game of football on a minor league team and eventually was scouted by the Philadelphia Eagles in the mid-70s where he was a wide receiver for three years. Papale is also a statistical anomaly because he is one of few professional football players, particularly nowadays, who did not play in college. Another (and more current) exception to this unspoken rule for playing in the NFL is Antonio Gates. Gates was a collegiate basketball player at my doctoral alma mater of Kent State University, and is currently a tight end for the San Diego Chargers. It's clear that neither Papale nor Gates were afforded the same tutelage in the game of football as the Manning boys, particularly since Papale's and Gates' four year hiatuses from football coincided with a time when they spent more time on the track and basketball court. It's safe to say that Papale and Gates are great examples of all-around athletes so would a brain scan reveal the same?

The last inspirational story in American football that I'd like to bring to your attention is that of Daniel "Rudy" Ruettiger. Although Rudy never made it to the NFL, he is one of few athletes to be carried off the field (in celebration) by one of America's most coveted football programs: the University of Notre Dame. Rudy's story is inspirational because he was undersized relative to most defensive ends, and he suffered from a learning disability: dyslexia. Further to Rudy's credit, his time with the Notre Dame football program was not through a traditional path. Prior to acceptance at Notre Dame, he spent six years serving in the Navy, working at a power plant, and studying nearby at Holy Cross College. At Notre Dame, Rudy was a walk-on. He worked hard, but did not actually see playing time until his senior year when he was thrown into the last few plays of Notre Dame's last game of the season against Georgia Tech. During these few plays, he had a sack, which inspired his teammates to the point that they carried him off the field. Unlike the Mannings', Vince Papale, and Antonio Gates, Rudy was no athletic freak with a future in the NFL. Instead, he impressed people and earned his right to play through mental toughness. And so I ask, what motivates someone like Rudy with a lack of size and experience to essentially go out and voluntarily be "thrown to the wolves" risking his life all for a few minutes of play? I hope to answer these questions (and more) in this chapter.

The Athletic Nervous System

By now, we should be comfortable with the idea that the brain is highly responsive and adaptive to exercise. In this chapter, I want to continue our discussion by pointing out unique differences in the anatomy of the brain between rookie versus elite athletes. In any case, let's begin with an extensive review of the brain structures that have been most often studied by neuroscientists and exercise physiologists. One important brain structure responsible for initiating a series of complicated movements is the cerebellum. This structure lies at the base of the brain by the brainstem.

The cerebellum is one of the few areas that has been preserved in structure and function across millions of years of evolution. It is found in all of our mammalian, reptilian, amphibian, and avian ancestors. In the human, there are roughly 3.6 times more nerve cells (neurons) in the cerebellum than in the superficial layers of the brain (cortex) even though the cerebellum makes up 10% of total brain volume! The cerebellum also receives feedback from the spinal cord in order to carry out movement.

One of the more exciting areas of research indicating that the spinal cord is essential for movement comes from the lab of one of my friends in graduate school; the lab of Dr. Jerry Silver at Case Western University in Cleveland, Ohio is using a radical approach to repair a severed spinal cord and restore movement by transfecting microbes sensitive to different colors of light in rodents. The experimental technique is known as optogenetics. Under this circumstance, scientists introduce a single-celled alga--*Chlamydomonas reinhardtii*--into the injured spinal cord. They then activate the microbes found on the spinal cord by shining a precise amount of blue light for specific intervals of time[3]. During this period of illumination, nerve cells in the spinal cord responsible for movement re-establish connections with other nerve cells. In some instances, the nerve cells continue to re-establish communication with other nerve cells even after the illumination is terminated. Over time, movement is eventually restored. In rodents, this process of restoration can take four days.

Perhaps one day this will be a viable option for restoring movement in humans. For now, one common approach used by physical therapists is to mobilize and strengthen paralyzed limbs. The intent is similar to that of optogenetics; introducing communication between the spinal cord and brain through routine stimulation. You can see some of these success stories at www.projectwalk.org. I have been fortunate enough to see this approach be somewhat successful in high school.

During my off-season for track, I used to do intensive plyometrics with thick rubber bands that provided different amounts of resistance. The program was called JumpStretch. The gym owner, Dick Hartzel, who is known in many high school, collegiate, and professional athletic communities as the “rubber band man”, patented the thick rubber bands. For years, Coach Hartzell had been working with a man who was largely immobile from the waist down. Through extensive and intensive physical therapy, the man was eventually able to slowly move his legs while upright although not entirely unassisted. Because I have not been back to Coach Hartzell’s gym in over a decade, I am unsure of this man’s progress, but at any rate, this story and supporting case studies do show how responsive and adaptive the spinal cord can be to routine stimulation.

Another brain structure important for the execution of movement is the pre-motor cortex which lies near the center of the brain in a deep crevice known as the central sulcus. There are many distinctive regions of the pre-motor cortex that are each responsible for controlling movement of a particular body part on a particular side of the body. Collectively, these regions make up the "homunculus" and have been mapped for function in animals by means of stimulating these regions with an electrified wire and observing movement. Further appreciation for the importance of the pre-motor cortex for the execution of movement derives from two experimental and clinical modalities.

First, we know that the pre-motor cortex is capable of being reorganized with experience by studying the brains of musicians and athletes with imaging technology like functional magnetic resonance imaging (fMRI). Depending on the type of skill required for a musician or athlete, whether it is hammering through a Mozart concerto on the piano or firing a soccer ball down the field, there is an increase in surface area in the region of the pre-motor cortex that corresponds with the predominant use of one's hands (pianist) or feet (soccer player) gained from practice. In fact, a neurologist could determine handedness in a professional musician or athlete simply by examining a brain scan of their pre-motor cortices and doing a size comparison[4].

In the clinical world, appreciation for the pre-motor cortex has been acquired through a phenomenon experienced in amputees known as phantom limb syndrome. Amputees often report of strange sensations emanating from the amputated limb that feel as if the limb is still there and fully functional. At the level of the brain, loss of a limb causes the region of the pre-motor cortex responsible for directing movement of this particular limb to remain, but nevertheless decrease in total surface area. To conclude, these clinical and experimental studies have shown that the pre-motor cortex abides by the general rule of thumb of fitness, "use it or lose it."

A third structure of the brain that deserves recognition in regards to movement is the frontal cortex. This area of the brain is unique to humans and other "intelligent" species like primates and dolphins. Unlike the cerebellum, the frontal cortex is a brain structure that has consistently increased in size and complexity throughout the course of evolution. A wide array of complex traits associated with consciousness, including decision-making, risk assessment, time management, and social skills, stem from the frontal cortex. These complex traits are also hallmarks of a successful quarterback.

Speaking of football, we know that the frontal cortex is responsible for all of these complex traits by studying the brains of retired professional football players pre- and post-death; many professional football players who have complained about being indecisive or emotionally volatile to a doctor have significant deteriorations of their frontal cortices. This condition--known as chronic traumatic encephalopathy (CTE)--results from concussions and repeated blows to the head[5].

In order to summarize how brain structures like the frontal and pre-motor cortices, the cerebellum, and the spinal cord are chronologically recruited and work together to execute a movement, I use an example of throwing a baseball. This example derives from an introductory neuroscience textbook that I read in college and is one that I use in the undergraduate neuroscience course that I teach[6]:

> *A seemingly straightforward skill like throwing a baseball requires a significant amount of coordination, timing, and energy at the level of the brain. The process begins with the frontal cortex having complete awareness of where the pitcher's body is in space (on the mound), time (bottom of the ninth), and situation (full count). Based on this initial information, a decision that is also derived from the frontal cortex must be made as whether to throw a fast, curve, or knuckle ball. Once this decision is made, the instructions for throwing a curve ball are relayed to the pre-motor cortex and cerebellum so as to recruit and activate the body parts required for throwing a curve ball. From here, the spinal cord is recruited and nerve cells along the muscles gate the precise range and timing of movement that the shoulder, elbow, hip, knee, and fingers much undergo for the curve ball to be successfully thrown.*

Now that I have reviewed the major structures of the brain that regulate movement, it is time to re-visit the scientific literature to document unique anatomical features and re-organization of the brain in an amateur compared with elite athlete. I must admit that a majority of the hits that I got on a widely used public database for biomedical research (PubMed) largely focused on the short- and long-term impacts of sports-related head injuries. So to clarify, this section will not focus on pathology, but will demonstrate to you how responsive and adaptive the brain can be with athletic training and experience.

"Why Athletes are Geniuses"

This is the subtitle from a *Discover Magazine* article written by scientific journalist, Carl Zimmer. The article described peculiar and advantageous differences in brain anatomy and function in elite athletes compared with amateurs[7]. I admire Zimmer's bold claim based on empirical evidence, of course, because of how often the terms "meathead" and "rocks for jocks" in relation to a college curriculum are thrown around. Some educational pundits have complained about how the academic records of athletes recruited to elite institutions with athletic clout (e.g. Duke, Stanford, Northwestern, and The Ivy League) are travesties to admission averages[8] and skew a college's ranking in *US World News & Report*.

I was treated no differently by the general student body as a *student*-athlete competing in The Ivy League conference, which is comprised of Brown, Columbia, Cornell, Dartmouth, Harvard, Princeton, University of Pennsylvania, and Yale. Ironically enough, many of the Fortune 500's CEOs overseeing our country's automobile

(General Motors), money market (Bank of America), food (Whole Foods), and technology (GE Companies) industries were also collegiate athletes[9]. But are these CEOs' successes a byproduct of access to significant academic and financial resources since many of them also played for The Ivy League, or could their "athlete" brains and (presumably) above-average social IQs and assessments of risks gained from athletic experiences have been the contributing factor?

In looking through the scientific literature, two types of methods are commonly used to compare brain anatomy and function in experienced versus amateur athletes. One approach is to use next-generation imaging technology like functional magnetic resonance imaging (fMRI) and positron emission tomography (PET). These technologies provide high-resolution 3D images through the use of strong magnets and radioactive tracers, which help to: 1) compare and contrast the size of specific brain areas; 2) to visualize the extent of connections between brain areas; 3) and to determine how metabolically active particular brain areas are.

The second approach is to place a vast array of electrodes around one's head in order to determine real-time changes in brain activity emitted from specific areas. This classic electrophysiological technique is known as electroencephalography (EEG), and has been used for almost a century to characterize brain activity during normal biological processes like sleep and learning and to monitor brain health following seizures or unconsciousness. EEG has been immensely refined across the last century. Until the late 1970s, scientists and doctors were limited by the ability to record brain activity over a few brain areas with about six electrical leads. Brain activity was observed via ink "scribbles" transcribed onto graph paper fed through a machine. Today, brain activity is recorded from hairnets with 256 electrical leads that are evenly spaced across the brain, and changes in brain activity in specific areas are digitally recorded and can be documented via 3D color-coded images.

Before I describe the unique changes in brain anatomy and function between an experienced versus amateur athlete, I must mention that these studies were not focused on a particular sport. Athletes from team sports such as volleyball, basketball, soccer, and badminton and individual sports such as diving have been studied. Many of these studies had overlapping findings.

First, experienced athletes had many brain regions that were bigger and thicker (i.e. had more surface area) compared with amateurs. The brain areas with greatest differences between experienced versus amateur athletes were important for communication (left temporal lobe of the cortex), spatial orientation (left parietal lobe of the cortex), risk-assessment, decision-making, and emotional reactivity (orbitofrontal cortex). Second, the experienced athletes had increased surface area of the cerebellum coupled with increased connections between the cerebellum and frontal and parietal areas of the cortex[10,11,12,13].

Moreover, similar to how our "homunculus" (pre-motor cortex) can change with athletic training and permanent injury, an increase in surface area devoted to communication, risk, on-the-spot decision-making, and spatial orientation would make sense for experienced athletes since these are fundamental skills gained from athletics. To summarize, these studies show that the brain of an athlete can become more highly developed with practice; by increasing space in certain brain areas devoted to a particular task and the number of connections necessary to execute a particular task. Now what about brain activity? Do these increases in size and connectivity found in experienced athletes require more energy and attention compared with that of an amateur?

As a high school runner, I often wore shirts that documented the "mind over matter" phenomenon characteristic of competitive long distance running; the idea that a person has to distract oneself or block fatigue and pain derived from oxygen deficiency and the accumulation of lactic acid from being consciously recognized. The "mind over matter" phenomenon may be necessary for twenty minutes, hours, or even days as is the case for the tribe of ultramarathon runners featured in Christopher McDougall's book *Born to Run: A Hidden Tribe, Super Athletes, and the Greatest Race the World Has Never Seen.* One of the most memorable shirts that I wore for high school track said "running is 10 percent physical, 90 percent mental, making us all insane!" on the back. While I wasn't a fan of this slogan and preferred others like "pain is only weakness leaving the body," is it really true that the brain activity and connectivity of a experienced athlete are different, and possibly heightened, compared with an amateur?

Even before competition or practice begins, the brain of an athlete rests more efficiently; compared with non-athletes, elite karate champions and fencers had a greater quantity of alpha waves, which are distinctive electrical spikes more often appearing across frontal and parietal regions of the brain[14]. Alpha waves are indicative of a resting state, and within my own field, are largely present during lighter stages of sleep (stage 2). They are also correlated with the extent of memory recall following a nap or night of sleep[15].

To quote Carl Zimmer who discussed this study on brain activity in karate champions and fencers, "an athlete's brain is like a race car idling in neutral, ready to spring into action." Even when the athlete brain is "sprung into action" during practice or competition, the extent of brain activity emitted from regions that regulate motor activity is markedly lower in athletes versus non-athletes, and it is a unique phenomenon that continues to evolve with practice[16]. This has been shown in groups of soccer athletes. Thus, practice does indeed make perfect.

The calm before the storm: My teammate and I waiting to drag a 400-pound sled across a soccer field at the 2013 Reebok Crossfit Games after some technical difficulties with the start signal. Photo by Ellie Toutant-Hoover.

The beauty of this brain plasticity acquired through athletic experience is that it allows the brain to focus on other aspects of the game that require more than just foot skill as is the case for soccer athletes. The potential of an athlete to effectively multi-task has been shown in one study of elite versus mediocre soccer players who were asked to dribble a ball through an obstacle course while simultaneously attending to a memory test of recognizing objects[7]. As expected, the elite players performed much better on the motor (dribbling) and memory (object recognition) tasks with little error compared with the mediocre players. Further, the amateurs could only perform well, relatively speaking, on the obstacle course for ball dribbling when it was not paired with the memory task. At the level of the brain, these performance "deficits" in mediocre players were linked to increased activity in the prefrontal cortex which regulates many of the complex traits that I discussed earlier—risk, decision-making, and spatial orientation, to name a few. For now, let's save additional discussion on the disconnect between mind, body, and performance for the next chapter when I discuss a phenomenon known as muscle memory.

Nevertheless, there may be a way for a mediocre soccer player to improve his motor and memory skill sets, although probably not to the level of Cristiano Ronaldo. This would require noninvasive stimulation of the brain. Many of you may have heard of deep brain stimulation, which has become an exciting research topic because of its effectiveness for managing depression, schizophrenia, and Parkinson's[17]. With deep brain stimulation, pacemakers are surgically implanted in brain areas that regulate certain psychiatric or neurological conditions, such as the striatum of the forebrain. These pacemakers act to re-stabilize electrical activity and control the extent of neurochemical release from the brain area.

Although we are far from justifying deep brain stimulation for the enhancement of athletic performance, researchers from Johns Hopkins University found great improvement in performance on a computer game, after controlling for time spent learning, in players whose scalp above the motor cortex received a battery jump across several days[18]. Whether or not this improvement in motor skills translates into athleticism is to be determined, but I would not be surprised if some "Ivan Drago" training team is experimenting with this technology.

Brain of a Sports Junkie

There are thousands of individuals in the US who annually spend thousands of dollars supporting some professional sports team or city of sports like Boston or New York. I always found this behavior strange, although I may be biased because I have never had to spend money on athletic apparel, practice and competition gear, and game-day travel throughout my years of high school, collegiate, and semi-professional athletics. Allegiance to a sports team goes well beyond purchasing a $100 jersey of the hottest recruit or spending $200 watching him live though. The Super Bowl is a great example. I traveled with friends from Atlanta to New Orleans to tailgate at Super Bowl XLVII (stadium blackout coupled with the retirement of Ray Lewis).

Since most of us were "professional students," we economized by renting an RV and camping spot, split the gas, and brought our own food. Factoring in libations on Bourbon Streets and local food fare, I spent close to $300 on the weekend. That's enough to raise some eyebrows, but not after spending many hours with our RV park neighbors who travelled in RVs equipped with 49ers or Raven-branded banners, inflatable decorations, and other NFL-branded accouterments from as far as California and Texas to spend the week reveling in pre-game recognition parties and attending the game. Many of our RV park neighbors admitted with little coercion that they spent $10,000-20,000 on that week!! So, what is it about the sports junkie's brain that entices him (or her) to spend a significant portion of their disposable income on something so intangible?

As a researcher in the field of drug addiction, I would argue that there is some unbalance in the sports junkie's reward circuitry that would entice them to make this arguably poor economic decision. In fact, the reward circuitry is a major biological substrate of study in a budding new area of research called neuroeconomics. So let's look at differences in brain activity in loyal fans of a winning versus losing team in America's second most anticipated sporting event of the year--March Madness—and America's favorite teams--Duke University Blue Devils and the University of North Carolina Tarheels.

In this study, researchers at Duke compared brain activity with fMRI neuroimaging in fans who were asked to watch a historical game between Duke and UNC wherein Duke won and UNC lost. I guess it is no surprise that the funding institution would choose this outcome.[19] The most important detail of this study is that every research subject had great knowledge of the Duke and UNC basketball programs; each person interested in participating in the study was pre-screened and asked to recall specific passes, shots, rebounds, and penalties of this historic game.

After this pre-screening, like-minded fans were asked to watch the game and then undergo fMRI testing. Most fans had very high levels of activity in the frontal cortex that regulate attention (anterior cingulate and dorsolateral cortices), areas in the temporal cortex that regulate memory formation and recall (medial temporal lobe), and motor areas like the basal ganglia, demonstrating that these die-hard fans were very much involved in the game, cognitively, emotionally, and behaviorally. However, Duke fans did have faster recall of key game moments compared with UNC fans. Keep in mind that Duke won this game.

Not surprisingly, there was a clear gender difference in regards to statistical knowledge about the game during the pre-screening period, the extent of recall of key game moments, and brain activity in motor areas. Thus, males were easier to recruit to the study compared with females and "performed better" in the experiment. We do not know whether these neurological traits of fandom and namely gender differences in fandom reported in this Duke study have a biological manifestation or are solely driven by society.

I guess an appropriate control would be to use this same testing paradigm for a memorable moment in women's gymnastics—"The Magnificent Seven" of the 1996 Atlanta Summer Olympics—or figure skating showdown between Nancy Kerrigan and Tonya Harding during the 1994 Lillehammer Winter Olympics. At any rate, it does appear that the brain of a sports junkie, behaves uniquely in the face of athletics much like that of an amateur athlete with practice or seasoned athlete.

Motivation: Inspired by Dopamine

I would like to continue this chapter with a discussion relevant to sports psychology because many of you may still not be convinced that one's potential to thrive on the community flag football team or in the New Orleans Superdome is a byproduct of environment coupled with hard-wired changes in the brain. As a lifelong athlete, my underlying motivation and subsequent training program for pushing myself to new physiological limits has varied greatly. In high school, I wanted to be state champion in several track and field events. In college, I wanted to be the best or second best in my events in order to compete in conference and other special meets that required cross-country and international travel. In graduate school and prior to finding Crossfit, I did not want to acquire cellulite, replace muscle with fat, or deal with any psychological anxiety common in retired athletes by continuing to train and compete in track and field competitions open to all ages. Nowadays, I want to be amongst the "fittest in the world" through Crossfit.

With changes in motivation have come changes in inspiration. For a brief period of time in high school, I was driven by an idea that I was fully utilizing an athletic gift from God. This was, ironically enough, followed by a period of craving for public recognition. In college, I was driven by my teammates, coaches, and the eyes of the football and hockey players in the weight room. In Crossfit, I am driven by the daily opportunity to compete for the best time on a workout and the euphoria manifest from oxygen deprivation.

But do my outlets for motivation and inspiration vary from other athletes and amateurs?

You'll notice that none of my motivations or inspirations for training emanates from a desire to be "like Mike [Jordan]" or to please my parents, which as I discussed early in the chapter, can be detrimental to one's athletic performance[1] if this parental pressure is overbearing. Worldwide and highly televised competitions like the Olympics and the World Cup can also motivate and inspire people to get off the couch or to keep working towards their athletic goals. Anecdotally, my decision to pursue gymnastics in 1996 was inspired by the "Magnificent Seven." Marion Jones' "record"-breaking performances in 2000 also pushed me through a grueling summer of training on the track.

After hosting the most recent 2012 Summer Olympics, over 750,000 more Brits began participating in some sport one or more times a week with women showing more interest than men and sports like cycling, swimming, and tennis seeing the largest increases in participation[20]. The US sees similar increases in soccer playing after the men's World Cup[21]. Whatever your motivation or inspiration may be for participating in sports, does the brain motivate us or is motivation socially driven?

Assessing the biology of motivation is much simpler in a laboratory animal than in a human, and it is actually something that I studied full-time in graduate school. In the field of drug addiction, scientists have identified brain areas, neurochemicals, and procedures over the past decade that are responsible for motivating an animal to self-administer drugs, even hard drugs like cocaine and heroin, directly to their brains[22]. Under these circumstances, the animal works for a drug by pressing a lever for a certain amount of time. This is learned through practice.

After the criteria (i.e. lever presses) for receiving a drug are met, the drug is then administered to one of the addiction-relevant areas of the brain. The most common brain areas that have been targeted in these experiments are those of the mesolimbic system, including the nucleus accumbens in the front of the brain and the ventral tegmental area of the brainstem. From these experiments, we also know that the neurochemical dopamine controls how much and how often the animal will self-administer the drug sometimes to the point of death. Leading up to self-administration, neurons that release dopamine fire rapidly[23] and pharmaceuticals that block dopamine release or that prevent dopamine from acting on neighboring neurons can stop the animal from self-administering more of the drug[22]. Obviously, in humans, this is one exploratory avenue for discovering effective treatments for drug addiction.

If dopamine is so important for seeking out hard drugs like cocaine and heroin, does dopamine also drive willingness to exercise? The impact of dopamine on exercise intensity has been widely studied in animals that love to run like mice and hamsters. As I discussed in the first chapter, my colleagues and I have found that the hamster's innate pleasure for running is so strong that it is a desired substitute for another pleasurable reward like alcohol. But we never looked at changes in dopamine in the brain.

Fortunately, other neuroscientists have found that animals with a love for running have huge increases in dopamine release from neurons as well as huge increases in the number of sites that dopamine can bind to on neighboring neurons during exercise regimens[24]. So in humans, are these changes in dopamine the neurobiological features that distinguish a couch potato from a seasoned athlete? But before we compare dopamine signaling in a couch potato versus seasoned athlete, we must determine if dopamine does drive how hard we are willing to work for something even if the payoff is negligible.

This question has recently been investigated in a neuroscience lab using everyone's preferred payoff: money. In this study, participants were given radioactive tracers, including a form of amphetamine, in order to activate brain centers releasing dopamine and to identify areas with increased activity. This approach is referred to as positron emission tomography (PET) imaging[25]. Following this prep phase of loading someone with radioactive tracers, the participants were asked to press a button for a certain period of time that would, in return, be associated with a certain payout at a particular probability; the monotony of this task is very similar to the drug self-administration studies in rodents.

After many rounds of testing, it was found that people with higher levels of dopamine were more willing to work for a big reward even if the likelihood of receiving it was low. This experimental circumstance is very reminiscent of statements made by the NCAA (National Collegiate Athletic Association) about the common fate of our country's collegiate athletes; "there are over 400,000 NCAA student-athletes, and just about all of [them] will be going pro, in something other than sports," which are then followed by Enterprise commercials that brag about being the largest employer of former collegiate athletes. So what is the point of sacrificing schoolwork, friendships, and relationships for twenty hours of weekly practice if the payout will likely not be a professional sports contract?

The biological answer relevant to dopamine is lifelong protection from aging-related declines in motor skills, improved cardiovascular and renal health, and better control of food intake [26]. Studies have found that enhancements in dopamine signaling in trained athletes versus sedentary individuals range from increased production of dopamine in the nerve cell, amplified release from the nerve terminal, reduced degradation of dopamine between nerve cells (referred to as the synapse), and increases in the number of sites that dopamine can bind to on nerve cells in the seasoned athlete versus couch potato.

Yes, these changes in dopamine signaling in the brain in trained athletes are similar to what have been observed in rodents with access to running wheels and likely explain the differences in one's motivation to exercise at the level of the brain. It is also no surprise that seasoned athletes have enhancements in dopamine signaling within motor areas of the brain like the striatum which dictates the precise timing and strength of movement with training.

As for long-term health, fine-tuned dopamine signaling manifest from long-term exercise protects against

Parkinson's and other aging-related declines in movement. As I discussed in the first chapter, movement disorders such as Parkinson's arise from a loss of dopamine signaling in the striatum. But this neuroprotection is only true to an extent. I say to an extent because one of the most beloved athletes of the twentieth century suffers from Parkinson's: Mohammed Ali. However, we do not know if Ali's condition is predisposed, meaning that his biological makeup conferred an increased likelihood of having Parkinson's, or if his Parkinson's is a byproduct of his violent sport. We must also consider race.

Contrary to the case of Mohammed Ali, it has been discovered that dopamine systems of the brain are more responsive to training in elite African endurance athletes compared with non-African competitors. One study even argues that this enhancement in dopamine signaling underlies African dominance in distance running[26]. In brief, the African athletes were reported to have increases in the production of dopamine, reductions in its degradation, and amplified calcium levels which assist in the release of dopamine from nerve cells relative to their non-African counterparts.

However, these studies did not control for biomechanical factors such as running gait or whether the athletes ran barefoot or wore shoes[27]; as discussed in extensive detail in Christopher McDougal's book *Born to Run: A Hidden Tribe, Super Athletes, and the Greatest Race the World Has Never Seen* running on the balls of one's feet, which is facilitated by running barefoot, is very efficient, especially if you plan on running 50 miles or more a day like the "hidden tribe of super athletes." Running barefoot also reduces the risk of skeletal and muscular injuries several folds. To summarize, we cannot conclusively say that dopamine signaling is responsible for the African dominance in distance running as there are many other factors such as running gait to consider, but I hope someone revisits this research area soon.

Welcome to the Pain Cave

I spend nearly an hour a day in physiological pain that is often comprised of profuse sweating, shortness of breath, burning in my lungs, nausea, muscle aches, and muscular fatigue to the point of failure. The coach of my competitive Crossfit team refers to this cascade of physiological phenomena as "entering a pain cave" because the onset of physiological pain is rapid and maintained sometimes for up to thirty minutes. This is especially true for workouts that are done in honor of a fallen soldier in order to put his or her suffering into perspective. In most cases after my workouts are completed, the pain subsides within five minutes, but how are my teammates and I able to push ourselves to even enter and stay in the pain cave every day? Instead of focusing on differences in one's strength, cardiovascular health, and ability to clear lactic acid from the body (i.e. what is responsible for that burning sensation in your lungs, gastrointestinal tract, and muscles during vigorous exercise) that are often driven by training and one's biological makeup, how does the brain respond and adapt to this physiological pain?

In the case of Crossfit, exceling in the pain cave can determine who advances to the Reebok Crossfit Games even when physical attributes are matched. To paraphrase one of the "fittest men on earth" and 2008 winner of the Games, Jason Khalipa, he overcomes pain and fatigue in a workout by *focusing on his technique, pretending to be coaching himself.* I was intrigued to hear this because I also engage in this same type of hallucinatory behavior, and will also become obsessive about my repetition scheme, counting in sets of 3, 5, and 10 throughout the course of completing 15-50 repetitions in an attempt to not put down the barbell or drop from the pull-up bar. I doubt that Jason and I are anomalies and that there are plenty of other Crossfit and non- Crossfit athletes with insane tolerance for the pain cave. Although the driving factors of pain tolerance in competition are fairly self-explanatory—beating an opponent (individual or team) at all costs or not appearing weak--what happens in the brain when one enters the pain cave?

The brain has its own type of pain relievers. They are called endogenous opioids, and they are much more powerful than ibuprofen or naproxen. Any athlete who has suffered a serious muscular injury (6-7 times over for me) can relate to the power of endogenous opioids because the extent of damage and pain associated with the injury may not be evident for a few hours or the next day after competition, which really sucks. The ability of competition to mask pain has also been studied in the lab. Researchers have recruited experienced and amateur male and female fencers, basketball players, and runners to compare their perception of physical and thermal pain during competition.

First, all of the experienced athletes got emotionally charged for competition. The amateurs did not. Across sports, the track and basketball athletes had higher tolerances for pain or less awareness of pain than fencers during competition. So does wearing a full-body suit for fencing have any bearing on these results or do the sports of track and basketball just attract more mentally tough athletes? The latter explanation seems to hold true in this study--that track and basketball athletes are mentally tougher versus fencers--because there were no differences in pain tolerance pre- and post-competition between each type of experienced athlete (runner vs. b-baller vs. fencer) and the amateurs.[28]

A quintessential post-workout pain pose at the 2013 Reebok Crossfit Southeast Regionals: "Jackie" is a Crossfit® workout

that locally hurts while it is being completed (i.e. lung and muscle burn) but globally hurts once complete, hence the hunched over position. Photo by Ellie Toutant-Hoover.

There are four different classes of opioids produced by the brain that are capable of masking all sorts of pain pre-, mid-, and post-competition. Endorphins are the most commonly recognized class; anaerobic athletes like sprinters and weightlifters will talk about the pre-competition "endorphin rush," while endurance athletes embrace the mid-competition "endorphin rush." Under both circumstances, endorphins mask pain for the short or long haul through their production and release in a subsection of the hypothalamus known as the arcuate nucleus.

The arcuate nucleus also controls hunger, satiety, and food intake. This is part of the reason why athletes, including myself, have to force food down their throats during a day of competition or have an immediate cessation of hunger pains mid-competition. The other classes of endogenous opioids include enkephalins, dynorphins, and endomorphins, which are released from the spinal cord, other subsections of the hypothalamus, and immune cells. Collectively, endogenous opioids mask the perception and physiological manifestations of pain, but can have short- and long-term side effects including constipation, nausea, drowsiness, and for female athletes, an irregular menstrual cycle[29]. I will return to a discussion on endogenous opioids in the last chapter of this book when I discuss tactics for enhancing athletic performance (i.e. doping) by manipulating the central nervous system.

As I conclude this chapter, I hope that many of you have overcome your "dumb jock" or "meathead" complex. The athlete brain can become far more advanced in structure and function with continuous practice and training. In the future, it would be worthwhile if neuroscientists would capture the brains of well-known athletes, particularly an entire bloodline of athletes like the Mannings' (Archie, Eli, and Peyton), to determine how heritable athleticism is.

Chapter 3

The Athlete Brain in Competition

Whether you are an athlete or a spectator, have you ever wondered how athletes are able to ignore cheers and jeers emanating from the stands and focus on the game at hand? Have you ever wondered if professional athletes get as many butterflies as amateurs before competition? Finally, as most fans in the city of Cleveland, Ohio can relate to, have you ever wondered why well-trained or professional athletes falter under the pressure? The previous chapter explored unique and adaptive changes in the athlete brain with practice and training. This chapter explores real-time changes in the athlete brain in the midst of competition. First, we will explore the psychological and physiological dynamics of a team sport versus individual athlete. From here, we will dissect the biology of game day, starting with pre-competition nerves and continuing with a discussion of muscle memory, overthinking, and choking. By the end of the chapter, the significance of practice and training should be evident because there is nothing much else one can do on game day except be bold and fight.

To Go Team or Go Individual

This psyche of a team athlete has always fascinated me. Since childhood, I have participated in both team and individual sports; basketball, softball, gymnastics, cross country, and track and field. By high school, I quit basketball and softball and became a "four seasons" (fall, winter, spring, and summer) athlete in gymnastics, cross country, and track and field, which are considered to be individual sports. Yet I would argue that gymnastics, cross country, and track and field are team sports to some degree; while every individuals' performance is measured, objectively for cross country and track and field and subjectively for gymnastics, the top two to five performances are factored into a team score to decide a winner. Therefore, I have never felt like a true individual athlete.

Regardless of my opinion, studies in the field of sports psychology have found numerous differences in the personality inventory of a "team" versus "individual" athlete. To avoid further confusion as I discuss these studies, a team athlete is someone who collectively and simultaneously works with a group of people to achieve one goal: win or lose. An individual athlete is someone who gets his or her performance evaluated either objectively or subjectively alongside everyone else. Everyone is scored or timed and everyone is ranked accordingly.

Many of the studies that focused on the personality inventory of a team versus individual athlete relied on the Big Five Model. The Big Five Model is a psychological examination of the following: 1) openness to experience—whether someone approaches unfamiliar situations with curiosity or caution; 2) conscientiousness—whether someone is diligent or easy-going and organized or careless; 3) neuroticism—whether someone is overly sensitive and nervous or secure and confident; 4) agreeableness—whether someone is friendly and compassionate or highly critical and detached; and 5) extroversion—whether someone is outgoing and energetic or solitary and reserved. This is a common model to examine personality inventories yet studies in sports psychology cannot agree on what types of personalities pre-determine or define one as a team versus individual athlete. Perhaps this is an issue of skill—evaluating an amateur versus seasoned athlete.

For example, a study conducted of female college athletes in 1967 reported that individual athletes are more introverted than team athletes whereas a study of female college athletes in 1968 reported that individual athletes are more extroverted than team athletes[1,2]. Independent of this discrepancy, a handful of other studies using the Big Five Model have found that team athletes are more generous and insightful whereas individual athletes are more dominant and aggressive. Of greatest significance, a majority of these studies found that athletes—regardless of being an individual versus team athlete—were more persevering, conscientious, extroverted, and aggressive than a population of non-athletes.

While the neurological and physiological mechanisms that underlie an athlete's personality did not come to fruition until years after these studies were completed, the neurological and physiological characteristics of an athlete versus non-athlete that were discussed in "The Athlete Brain" align with the general personality inventory of an athlete: 1) Elevated dopamine with training drives continued motivation to train; 2) Elevated testosterone with training drives aggression; and 3) Elevated serotonin with training drives social interaction.

The elation of competing and winning on a team: Science reveals that both team and individual sport athletes are uniquely and psychologically different from non-athletes when personality inventories are considered. Photo by Ellie Toutant-Hoover.

Another finding that these studies of personality inventory did agree upon is that team and individual athletes are equally as emotionally intelligent and, more importantly, both groups are more emotionally intelligent than non-athletes. At first, I found these results surprising because "team work", "selflessness," and "bonding" are buzzwords in the world of team and not individual sports that infer some level of emotional connection or compromise for the greater good. Rarely do you hear these buzzwords to describe an individual athlete. Regardless of my opinion, what is emotional intelligence and why at a biological level are athletes more (emotionally) intelligent?

Emotional intelligence is a measure of how easily and accurately one is able to recognize, analyze, and regulate their own and others emotional states. Much like how general intelligence is measured with an IQ test, there are validated "EQ" (emotional quotient) tests to measure emotional intelligence. There are other means to measure emotional intelligence that have not been widely used to study athletes, but hopefully this will be *a priori* soon. In the meantime, a few colleagues of mine have studied the impact of sleep, particularly napping, on emotional intelligence in healthy young adults[3].

In their study, the research subjects who were primarily college students at University of California-Berkeley were asked to get adequate sleep under their normal sleep-wake schedule before pulling an all-nighter that forced the subjects to stay awake for 24 hours. During this all-nighter, the subjects were shown a series of images that were aimed to evoke negative emotional responses such as fear, anger, and envy. These emotionally salient images were flashed alongside images that were reliably known to evoke ambiguity (could be good or bad) or no emotional response. During this time, a brain scan was also undertaken using functional magnetic resonance imaging (fMRI). The scan was used to probe for changes in brain activity in areas that regulate emotion, attention, and sleep. As expected, the degree of sleep deprivation was positively correlated with the degree to which the subjects over-reacted to all of the images: negative, ambiguous, or neutral.

In looking over the brain scans, a majority of this emotional over-reactivity was due to heightened activity in the amygdala—a major structure of the limbic system that I discussed in "Barbells and Brains"—as well as in many regions across the cortex. From the perspective of survival in the wild, these results make sense. Inadvertent sleep deprivation, muscle fatigue, mental fogginess, and hunger make an organism more vulnerable to attack. Therefore, it makes sense in the face of survival that the brain would overreact to any stimulus whether it's a serious threat or benign.

Part of me wonders if the student-athletes of UC-Berkeley would be less emotionally volatile in this study if it is true that college athletes are more emotionally intelligent and that athletics and fitness fine-tune the "go"/"no-go" circuitry of our brain responsible for controlling emotions as I discussed in "Barbells and Brains". Further, because competitive athletes are well familiar with muscle fatigue, mental fogginess, and hunger during training and competition, would they be better able to adapt to the sleep deprivation and therefore be less emotionally volatile in this study?

At the other end of the spectrum, part of me also wonders if sleep deprivation is responsible for instances of "hotheadedness" and "roid rage" on the football field or in the hockey rink; this is a legitimate question since athletes often have poor sleep on the night prior to competition coupled with the fact that a significant proportion of collegiate and professional football players suffer from sleep apnea due to their size[4]. Perhaps this is the biological explanation for why the two most aggressive and emotionally volatile defensive players in the NFL today—Ndamukong Suh and James Harrison (who hails from my doctoral alma mater of Kent State University)—are repeated offenders of unsafe tackles and poor sportsmanship. At the very least, the Berkeley study points to the benefits of taking a nap with a high volume of training and competition. It very may well result in less yellow cards and flags thrown for poor sportsmanship.

Brain of a Daredevil

After watching several summer and winter Olympics, it is clearly evident that some sports are more dangerous than others. Many people would consider winter sport athletes to be psychotic. That is, what type of courage does it take to cram four adult males into a 1,400-pound hunk of metal that corkscrews down ice at 80 mph? Or what is the mental preparation involved for doing a double back tuck over and then re-catching a thin metal bar suspended 10 feet above the ground? Moreover, is any athlete capable of overlooking serious injury or death for the enjoyment of sport or is the daredevil trait reserved for a unique group of biological mutants?

Although I do not bobsled, ski black diamonds, or do the half-pipe, I did gymnastics for nearly two decades and still pole vault to this day. I have had several near death experiences as a result of falling from the high bar or from a pole wavering above a metal box while being upside down and 13 feet above the ground. These near death experiences did not just happen once for gymnastics and once for pole vault, but rather on several occasions for each. In both cases, did I stay away from the uneven bars or refuse to pick up a vaulting pole for weeks after the accident? Nope. Instead, I mildly enjoyed the brief moment of panic and flood of chemicals pumped from my adrenals, shrugged it off, and went back to my starting position. This is the common psychological mindset of an athlete competing in a dangerous sport. But what about the physiological mind set of these athletes?

The thrill of pole vaulting: Words cannot describe the physiological euphoria associated with hanging from a pole, being launched over ten feet in the air, and falling onto a soft mat on the back end, on most occasions. Photo credit by Dan Grossman

Fortunately, neuroscience cares about athletic daredevils. The biological factor and neurochemical that has received the most attention should not be a surprise by now--dopamine. As I discussed in "Barbells and Brains" and "The Athlete Brain," dopamine is the neurochemical of positive traits like pleasure and motivation and simultaneously negative traits like pain and addiction. Of the numerous studies that I examined, athletic daredevils exist because of an imbalance or altered set point in dopamine. The dynamics of dopamine and how it determines one's desire and passion to take athletic risks is governed by two basic physiological principles: homeostasis and allostasis. In order to understand allostasis, you must first understand homeostasis. Homeostasis refers to the set point of any physiological process. It is commonly used in reference to our core body temperature or hormone profiles. If there is some deviation in a physiological set point, the body and brain will adapt accordingly in order to re-achieve that set point. The best example is shivering to stay warm when core body temperature drops in cold weather or sweating to cool down when core body temperature rises during a workout. Sometimes, however, a biological process has deviated too far from its physiological set point for too long forcing a new set point. This is the definition of allostasis. Drug addiction is postulated to arise from allostasis as is athletic thrill seeking.

The brain of a daredevil is rarely satisfied and always seeking a "dopamine pump." It is constantly in a state of allostasis. Thus, it is the real-time, act of fighting gravity, avoiding unmovable obstacles, and risking injury and death that sends the dopamine system of a daredevil into overdrive. Only then is the brain of the athletic daredevil satisfied, for the time being. As training progresses, the brain of an athletic daredevil requires a better and grander "dopamine pump," requiring more complex tricks and obstacles that subsequently present greater risks for injury and death. Over time, a new dopamine set point (allostasis) is achieved and the cycle begins anew. As far as the source, dopamine manifests from the forebrain, particularly in the substantia nigra—which is the control center for gross and fine motor movements—and the nucleus accumbens—which regulates rewards of many sorts. It also is produced and released from other reward centers in the brainstem.

Beyond the release of dopamine, special attention has also been paid to genetic differences in athletic daredevils. This was most recently studied in a group of five hundred skiers and snowboarders. In addition to collecting the skiers' and snowboarders' DNA, the researchers gave them a series of questionnaires that examined impulsivity and sensation-seeking. The genetic variant that was most striking in the skiers and snowboarders and that most corresponded to risk-taking and sensation-seeking behaviors was the dopamine-4-receptor gene (DRD4). This gene encodes for one of five different receptors that dopamine can bind to[5].

After seeing this study, I immediately went back through my 23andMe profile—a popular and affordable genetic testing service—and found that I not only have several genetic variants encoding for dopamine, but also a high risk for disorders related to too much dopamine—restless legs and addiction. For me, these studies and my genetic landscape beg the question of whether the act of becoming a daredevil is "hard-wired" or a byproduct of environment; neither of my parents were athletes, but yet I have always sought out and enjoy risky sports. To justify their support for my athletic shenanigans, my parents still tell people to this day that I willingly jumped off the three-meter dive with no floaties when I was three years old. Much like the discussion of whether athleticism is "hard-wired," a byproduct of the environment, or impacted by both nature and nuture in "The Athlete Brain", we may never scientifically know about the rationale behind the athletic daredevil unless the interest and financial backing are there.

Pre-Game Jitters

While listening to my favorite sports talk radio program, *The Tony Kornheiser* show one day, Tony Kornheiser was interviewing some soon-to-be-retired athlete. I can't remember the guest's name or what sport he played, but something he said has stuck with me for years. Tony asked his guest how he knew that he was ready to mentally retire to which his guest replied, *because I don't get nervous for games anymore. I'm not sure if it's not caring or experience, but it's bothersome.*

The bottom line is that an athlete who wants to succeed, either individually or collectively as a team, will undoubtedly get nervous prior to competition. It is more or less a pre-game superstition much like wearing a certain pair of socks, eating certain foods, or listening to certain music during warm-ups. In fact, general game day anxiety is further worsened by forgetting one's pre-game superstition; it is difficult for athletes to migrate from their comfort zones of training as is, from performing on different equipment to adapting to a new climate.

Pre-game superstitions help athletes gain a sense of control over their performances even if nothing else is in their favor. Studies have shown that a healthy dose of habitual, repetitive behaviors much like a methodical pre-game warm-up or meal helps reduce anxiety not just in humans, but also across the animal kingdom[6]. A group of biologists speculate that habitual, repetitive behaviors help prepare an organism for actual, serious threats that require a means of escape or defense.

Pre-game jitters and superstitions are infamous to the world of professional baseball. As one professional baseball remarks in an *ESPN: The Magazine* article on baseball superstitions, "Players are superstitious about being superstitious.[7]" From swinging a particular bat in a certain manner before approaching the mound, to winding up at a particular angle and a certain number of times while on the mound, to chomping on a particular brand of tobacco or chewing gum while in the dugout, baseball athletes have plenty of eccentric superstitions. My hometown team---the

Atlanta Braves—are convinced that the "hits are in the watermelon [Supper Bubble™ flavored] gum," and not any other flavor[7].

Outside of baseball, I have been amazed to learn that Usain Bolt—the fastest and one of the fittest humans on Earth—ate McDonald's Chicken McNuggets every day in the dining hall during the 2008 Olympics. But he did not just eat two, but rather enough to sustain his daily caloric load—about 100—totaling about 47,000 calories over the course of the 2008 Olympics in Beijing, China.

Beyond just identifying that pre-game superstitions help reduce general anxiety on game day, what are the biological mechanisms, particularly at the level of the brain, behind game day anxiety? There are two systems that I would like to discuss of which you have been thoroughly introduced to in "Barbells and Brains" and "The Athlete Brain." These two systems are the hypothalamic-pituitary-adrenal (HPA) axis and the limbic system. The HPA axis is a broad system with structures and chemical cascades in the brain—the hypothalamus and pituitary—and in the periphery—the adrenal glands. The release of hormones, particularly cortisol from the adrenals, helps to prepare the body and brain to fight or flight. It is activated immediately in the presence of stress, but it can also be recruited in anticipation of stress.

The greatest benefit of the HPA axis for athletics is that it provides the necessary energy reserves for competition. Glycogen stored in the liver is broken down into glucose that is then transported to the brain and muscle as fuel. Fats, which are a sustained source of energy, are also made available. The problem with pre-mature activation of the HPA axis prior to game-day is that it often prevents a night of well-rested sleep. This is why I always advise my athletes to get adequate sleep across the week leading up to a big competition and not just the night before.

The limbic system also contributes to pre-game jitters. Unlike the HPA axis, the limbic system is strictly present in the brain. The two major structures of the limbic system are the amygdala and the hippocampus, but the HPA axis and the limbic system do interact. The hippocampus can turn off the HPA axis by inhibiting the release of corticotropin-releasing hormone (CRH) from the paraventricular nucleus (PVN) of the hypothalamus. This "go/no go" signal arising from the limbic system onto the HPA axis works well with intermittent stress, but it begins to falter with long-term stress. As a result, the HPA axis stays active and the hippocampus begins to degenerate.

Both types of pathology—sustained activation of the HPA axis and degeneration of the hippocampus---have been shown on numerous occasions in animal models. Whether or not decades of game stress causes similar demise to the HPA axis and limbic system is unknown, but perhaps the recent effort to dissect brain structure and function in deceased athletes who have donated their brains to science and medicine will shed light on this research area.

Aside from contributing to pre-game jitters, the limbic system is also responsible for helping us remember moments of joy and pain in a single game. In "Barbells and Brains," I discussed emotional memory, which is the process through which a strong emotion is tied to a particular event or object. A football player who runs into the end zone for the first time in his collegiate career or to the contrary, badly sprains his ankle during the last game of the season will remember the exact moments, sounds, and smells leading up to and after this touchdown or injury for the remainder of his life. I recall the first day of competition in the Reebok Crossfit Games with absolute clarity, and I can also recall the day that I blew out my back after falling twelve feet onto the pole vault crossbar with the same level of clarity.

Both structures of the limbic system—the amygdala and hippocampus—are responsible for this clarity. The amygdala provides the emotion to be felt—elation, disappointment, or depression—while the hippocampus processes (encodes) this information for later retrieval. Further, perhaps this mechanism of processing and retrieval by the limbic system is another process through which pre-game jitters arise. Regardless, professional athletes can assure you that being nervous for competition is perfectly normal. You should worry if you're not worried.

Muscle Memory

My event coach in college used to say, "the problem with coaching you is that you're too smart." There are times when athletes should shut their minds off. Some of the most historically missed catches of an uneven bar, football, or barbell often results from overthinking. I can relate. Some of the best days that I have ever had at gymnastics or pole vault practice were when I was mentally drained. In Crossfit, my Olympic lifts, which require a high degree of technical and motor skill, are better when I'm mentally fatigued. The reason for improved performance despite little conscious effort is due to a central pattern generator that lies deep in the brain. This central pattern generator arises and improves with practice. As Bruce Lee once said, "I fear not the man who has practiced 10,000 kicks once, but rather the man who has practiced one kick 10,000 times." The "10,000 rule" of practice has also been popularized through the nonfictional work of Malcolm Gladwell. But what is the central pattern generator and how can it dictate athletic success in the absence of conscious thought?

In athletics, the central pattern generator is referred to as muscle memory. In brief, muscle memory is the process through which the brain, the spinal cord, and muscle and tendons communicate with each other in order to encode and remember how to execute a skill. Muscle memory is the foundation of highly technical sports such as Olympic weightlifting and gymnastics. It is also the reason why we can learn to ride a bike when we are five years old and can still ride one twenty years later. Let's spend the remainder of this chapter discussing muscle memory at level of the nervous system—first the brain, then the spinal cord, and then special nerves along muscles and tendons.

Central pattern generator of performing the Olympic snatch: A highly skilled lift requiring the recruitment of many upper, mid, and lower body muscles in quick succession by the brain and spinal cord. Photo by Clayton McKenney of Niko Vandevoorde (Crossfit RX; Atlanta, GA).

The major circuit of muscle memory in the brain is the basal ganglia. The basal ganglia is not just one structure, but comprised of several structures that lie in the forebrain: 1) the nucleus accumbens, which is also a major component of the reward circuit; 2) the substantia nigra; 3) the caudate nucleus; 4) the putamen; 5) the globus pallidus; and 6) the subthalamic nucleus. These structures, both big and small, work together to coordinate gross movement of limbs and to calibrate fine movement of fingers and toes.

We know that this is true from the study of animal models as well as from the study of movement disorders such as Parkinson's, Huntington's, and even Tourette's in humans. In rodents, the removal of a specific structure or group of structures in the basal ganglia can tell us what each structure is responsible for; regulation of gross or fine movements as well as initiating or inhibiting movement. The administration of certain drugs within the basal ganglia has also provided insight on what neurochemical cascades are required for initiating, controlling, and stopping movement. Two of the most commonly studied neurochemical cascades are dopamine—an excitatory neurotransmitter—and GABA—an inhibitory neurotransmitter.

The major dopamine production centers of the basal ganglia are the nucleus accumbens and the substantia nigra. The substantia nigra has received significant attention in the study of movement disorders because destruction of the dopamine-producing cells in the substantia nigra is the main pathology responsible for Parkinson's disease: a movement disorder characterized by slow, uncoordinated, and shaky movements. Although Parkinson's is a disease of neurodegeneration, it is treatable to an extent with L-Dopa. In the world of biochemistry, L-Dopa is the first ingredient necessary for producing dopamine. In the world of medicine, L-Dopa sold under several trade names increases the concentration of dopamine in the brain.

Because exercise also increases dopamine concentrations in the brain, exercise is recommended for individuals with a family history of Parkinson's or individuals in the beginning stages of Parkinson's. One would think that athletes are at a reduced risk for Parkinson's. This is true to an extent, however, it is ironic that one of the fittest and most coordinated athletes in the history of sports—Muhammad Ali—suffers from Parkinson's. Nevertheless, a decade of blows to the head at high speeds is likely not to protect one's brain from destruction.

In contrast to Parkinson's, Huntington's is characterized by excessive hyperactivity, flamboyant dancing known as chorea, and an inability to control one's movement. It largely manifests from a genetic mutation that is highly heritable meaning that if one parent has Huntington's, then their child also has a 50% chance of having Huntington's. Beyond genetics, the two areas of the basal ganglia that are most commonly studied in the progression of the disease are the caudate nucleus and putamen. To manage Huntington's, patients are given drugs that inhibit the central nervous system such as benzodiazepines.

Because Huntington's is a disease of excessive movement and not a lack of movement, it is unclear if exercise is beneficial for managing Huntington's. From the perspective of mechanism, Parkinson's and Huntington's disease reveal the differing responsibilities of the substantia nigra and the caudate and putamen areas of the basal ganglia. Slow, uncoordinated, and shaky movements characteristic of Parkinson's suggest that the substantia nigra provides the "go" signal for movement, whereas uncontrollable dancing characteristic of Huntington's suggests that the caudate/ putamen provides the "no-go" signal or fine tuner of movement.

Now that I have covered the basic mechanisms of the basal ganglia, what is unique about the basal ganglia of athletes? Unfortunately, this has received little attention in the biomedical community. One of the few studies that exists on the subject matter examined mental rehearsal in elite versus rookie archers. Mental rehearsal is very much part of game day preparation. For me, this is usually why I can't sleep the night before a big competition; because I am obsessively strategizing and rehearsing my anticipated performance. In this study, elite and rookie archers were asked to mentally rehearse their performances while inside a functional magnetic resonance machine (fMRI). After scrutinizing the brain scans, the researchers found that the brains of the elite archers were more efficient; that is, less brain areas were active at a single moment in time compared with the rookie archers[8].

Where have we seen this before? In "The Athlete Brain," we learned that the brains of seasoned soccer players are more efficient because the seasoned soccer players were better able to multi-task and attend to an experimental test while dribbling a soccer ball through an obstacle course whereas rookies were incapable of attending to even one task successfully. When areas of activity in the brain of elite and rookie archers were compared, the basal ganglia of the rookie archers were more active. Essentially, this is neural evidence of learning, executing, and fine-tuning a complex motor skill in real-time. It would likely take 10,000 more shots before the brains of the rookies behaved more like those of elite archers, but that is what practice is for.

Outside of the brain, muscle memory requires the spinal cord, which is also characteristically part of the central nervous system. The spinal cord is divided into specific subsections with each subsection being responsible for movement across a particular region of the body; nerves originating from the cervical section control movements of the neck and upper torso, those of the thoracic region control movements of the mid- and lower torso, those of the lumbar region control movements of the upper legs, and those of the sacral region control movements of the lower legs. Within the spinal cord, the central pattern generators that are responsible for executing and fine-tuning movement arise from a unique group of nerve cells called interneurons.

Interneurons collect information from nerve cells dictating movement or nerve cells dictating sensations such as touch and pain that lie in muscles and tendons and then these interneurons transmit information about sensation and pain to the brain. This pathway can also work in reverse such that information from the brain about where and how to move in space and time is transmitted to the interneurons of the spinal cord. Finally, it is possible for communication to only occur between interneurons of the spinal cord and muscles and tendons. This is the neurological definition of a reflex of which we will further dissect in a few paragraphs.

The reason why interneurons are such an integral part of muscle memory is because they serve as the gateway of communication between muscles and the brain. Of greatest fascination, interneurons are able to amplify, stop, or fine-tune signals coming from the brain or muscles because of a unique neurological process called disinhibition. Disinhibition is the inhibition of an inhibitory nerve cell such as one that releases GABA. By inhibiting an inhibitory nerve cell, the next nerve cell in the chain of command becomes activated instead of inhibited like it normally is. Thus, the interneurons of the spinal cord are inhibitory. Therefore, when an inhibitory signal from the muscles or brain reaches the interneurons, the signal gets amplified instead of suppressed. This is one means through which a central pattern generator responsible for a complex series of movement can be calibrated or in some cases, cause someone to have a "mental block," bailing out of an Olympic lift, tumbling pass, or football throw.

Muscle memory would not be complete without a contribution of muscles and tendons. There are two distinctive groups of nerve fibers along the muscles and tendons called muscle spindles and Golgi tendon organs. As one can infer from name alone, muscle spindles lie along muscles, whereas Golgi tendon organs lie along tendons. A muscle spindle is responsible for alerting the spinal cord and brain about muscle tension and stretch (Side note: In this case, it is inaccurate to say that the muscle is "flexed" unless you know that the muscle is a flexor. There are categories of muscles that often lie on opposite ends of a limb: flexors and extensors. Flexors like the biceps shorten in space when activated, whereas extensors like the triceps elongate when activated). Sensory neurons also work alongside muscle spindles to ensure that the muscle is not over-flexed or over-stretched, which could result in short-term or permanent damage. This motor circuitry is also required for a basic reflex such as those often tested in the doctor's office. In fact, the circuitry of basic reflexes does not require the brain. It simply requires functional muscle spindles and sensory neurons of the muscles and interneurons of the spinal cord.

As far as Golgi tendon organs, they are referred to as proprioceptors because they tell the spinal cord and brain where a muscle or group of muscles is in space. Golgi tendon organs are responsible for knowing how deeply you have squatted without actually looking in a mirror. Collectively, muscle spindles and Golgi tendon organs tell athletes where they are in space and time without much conscious thought. The efforts of muscle spindles, Golgi tendon organs, and the entire concept of muscle memory are very well regarded in Olympic lifting circles. For example, a missed snatch can arise when an arm is not entirely straight or a knee is not entirely turned out. Missing the lift while in a compromised position is a neurological or neuroprotective means to avoid injury by overstretching of a muscle or tendon. Thus, the risk for injury skyrockets when the lifter attempts to save the lift while in a compromised position.

For strength and conditioning coaches, it is extremely important to take the mechanics of muscle spindles and Golgi tendon organs into consideration during an athletic season to avoid overtraining an athlete. The process through which muscle spindles and Golgi tendon organs become overworked is referred to as neurological fatigue. Under conditions of neurological fatigue, the muscles cannot generate as much force or work for their usual period of time, causing athletes to not lift as much weight as they normally do or to get injured in the process.

This is why rest and recovery is just as important to an Olympic or power lifter as training. This is also why many athletes will achieve a new personal best on an Olympic or power lift after a few days off. Fortunately, neurological fatigue does not last more than a week. Further, studies in power lifters and Olympic lifters have shown that men and women acquire and lose neurological fatigue at different rates. The time window wherein there is a decrease in performance due to too much rest is often shorter for females than males. Obviously, the biological implication of this study is that testosterone (or estrogen) is driving this difference since this is the one apparent physiological difference between men and women.

The Big Choke

As a lifetime fan of Cleveland sports, I have habituated to athletes choking. I call it "Charles Nagy Syndrome," named after the great pitcher of the Cleveland Indians who led the team to two American League pennants in 1995 and 1997, but could never clinch a World Series title for the city of Cleveland. A few years ago, I wrote an opinion editorial for Kent State University's student paper where I did graduate work, lamenting about how we continue to have loyal fandom despite routine disappointment from professional and collegiate athletes in Ohio; at that time, The Ohio State University Buckeyes lost yet another BCS national championship game to the Louisiana State University Tigers[9]. In looking over the history of Cleveland and Ohio sports, are these historical moments of choking just a fluke or is choking endemic and part of human nature?

For seasoned athletes, it is often easy to ignore distractions such as cheers, jeers, and the number of fans during a game. These distractions are likely to cause an amateur to choke, but not a seasoned athlete. However, this does not mean that performance under pressure is nonexistent for a seasoned athlete. From personal experience, I would say that the pressure to perform is higher the better you are. Further, a missed catch for a professional athlete can have dire consequences—socially, financially, and personally; although Wes Welker is a very good wide receiver in the NFL, his "butterfinger" catch as a New England Patriot during the final moments of Super Bowl XLVI will indefinitely haunt him, personally and publicly. Shortly after, he was traded to the Denver Broncos. To this end, what is the neurological mechanism of failing to perform under pressure?

Earlier in the chapter, I mentioned that overthinking could result in athletic failure. In that context, I brought up overthinking to demonstrate the beauty of movement circuitry; how the brain, spinal cord, and muscles and tendons can work together in the absence of conscious thought. In this section, I will provide neurobiological evidence for why overthinking can result in athletic failure. For more specifics on the neuroscience of choking, I recommend reading Dr. Sian Beilock's book, *Choke: What the Secrets of the Brain Reveal About Getting It Right When You Have To,* who is an expert on the subject matter. For now, I will just cover the basics.

Many of Dr. Beilock's studies focus on performance when the stakes are high; such as for money or when someone else is watching. In a notable study, Dr. Beilock had subjects come into her lab at the University of Chicago to putt on a green. Putting not only frustrated Billy Madison and has led to one golfer winning the Master's over another, but it can make a Sunday afternoon of miniature golf turn sour; putting is probably one of the most universally athletic skills where choking and overthinking are common. Like many neuroscience studies focused on performance, Dr. Beilock coupled a measure of behavioral performance—putting—with a measure of brain structure and function—an fMRI.

Through such studies of performance under pressure, Dr. Beilock's laboratory has found that both the basal ganglia and frontal cortex—two areas that I have routinely discussed as changing and adapting to training throughout the book—contribute to the science of choking. The general idea is that when muscle memory is high and cortical activity is low, in that the putter just putts the ball and does not overanalyze the angle, speed, and force at which the club hits the ball, then performance is usually better. However, when activity in the frontal cortex, the epicenter of decision-making and attention, is high, then performance begins to falter at a predictable rate[10]. Overthinking at the level of the frontal cortex and subsequent performance are even worse when there is a group of people to watch. To be trite, Dr. Beilock's research shows that there is some benefit to being a "dumb jock" and shutting one's mind off. Perhaps this is the rationale behind Nike's slogan, "Just Do It."

As I conclude this chapter, I hope that many of you have grown to accept the challenges athletes face on game day. Between managing general pre-game anxiety, partaking in an eccentric pre-game ritual, envisioning the perfect game, executing the perfect pitch, and performing under pressure, athletes must find balance between mind and body, or their brain, spinal cord, and muscles. At the end of the day, an athlete who trains and competes is better off than someone who does not. Their brain becomes more efficient and better able to adapt to stress from the experience. It's just best to not overthink.

Chapter 4

Train, Eat, Sleep

Have you ever done a night road race, either for enjoyment or competition? If so, you may have felt different—both physiologically and mentally—and may have even underperformed. This is because most people don't train at night. I can relate. Every year, my friend from high school and I run in the Disney Wine and Dine Half Marathon which starts around 10:00 p.m. Around 1:00 a.m., we reach the finish line, and are then shuttled to an after party at the World Showcase of Epcot that lasts until 4:00 a.m. I train for the half marathon every year, and I always end up running more than 13.1 miles at some point during my training. But it doesn't matter because I always feel physiologically emaciated every time I finish the race even if my racing pace is comparable to that during my training runs, and I properly hydrate and fuel along the course. Some may say that I choke on race day, but as my research shows, my body does not want to be moving let alone be running 13.1 miles at 1:00 a.m. This chapter will focus on my own area of research with some emphasis on animal models of exercise since the systems that I will discuss have been nicely conserved across the evolution of mammals. By the end, you will learn that there is a certain time to exercise and to perform well and a certain time to sleep (and get enough of it). You will learn that the body can be very stubborn against any change in this regimen. I will also summarize some research that is able to predict if an East or West Coast team will win based on whom they play and when.

A Body Full of Clocks

Have you ever maintained a strict work, eating, and sleep schedule to the point that you didn't need an alarm clock to wake up in the morning? Or, do you find yourself waking up with the sun regardless of how long you had slept or what time it is? If so, then you are a "slave" to your internal biological clocks (note: this analogy is commonly used in our field). A biological clock has been discovered in almost every tissue of the human (and mammalian) body, including the brain, the heart, the liver, muscle, and even fat and red blood cells[1, 2]. The "master" clock lies deep in the hypothalamus in a structure known as the suprachiasmatic nucleus. This structure is largely responsible for why you wake up before your alarm or rise with the sun. It is in direct communication with the visual system and has many lines of communication with the other biological clocks throughout the body[3].

These biological clocks are also highly responsive and adaptive to exercise, as we know from research done by the lab at Kent State University where I completed my PhD.

Most of this research has been done in nature's greatest ultra-marathoner: the hamster; hamsters not only love to run, but they become very moody and self-destructive to the point that they drink significant amounts of alcohol (when given access to it) if they are prevented from running[4]. This was very fascinating to study in graduate school, especially when my lab mates and I encountered hamsters that had run close to nine miles a day on their wheels. On average, our hamsters would voluntarily run about four miles a day. Some of them would run ten miles! This means that the average human male would have to run about thirty-six miles a day to keep up with hamsters. The Tamahumara Indians featured in Christopher McDougall's book *Born to Run: A Hidden Tribe, Super Athletes, and the Greatest Race the World Has Never Seen* are capable of this, but for the average American, I think not.

Allowing hamsters to run when they should be sleeping also shifts their biological clocks by a few hours[5]. My doctoral advisor, Dr. J David Glass, has spent nearly a decade trying to understand the mechanisms behind this shifting, and it appears that the neurochemical of exercise—serotonin--plays a major role. Serotonin interacts with other neurochemicals like neuropeptide Y released from a subsection of the thalamus, just above the hypothalamus. This neurochemical chain of events is necessary in order for shifting of the biological clock by exercise to occur.

We know this from administering pharmaceuticals that antagonize serotonin or neuropeptide Y and that subsequently block this exercise-driven shifting[6]. In addition to a single shift in the biological clock by exercise, having hamsters run at a certain time of day everyday causes the biological clocks and the many physiological and behavioral phenomena that are wired to these biological clocks to re-adjust and re-organize themselves around the time when the hamsters are about to run[5]. This is known as anticipatory activity.

Is this approach the billion-dollar solution to motivating the average human to run more? At the very least, this research points to why we should continue to exercise at a certain time of day or why athletes should train at the same time of day that they compete. Even small deviations in our well-constructed schedules of sleep and exercising can have dire consequences on performance and health; in fact, we do know that risks for heart attacks are greater on the day after daylight savings[7] due to disruption of the biological clocks in the brain and heart. I will return to the importance of exercising at a certain time of the day later in the chapter when I talk about how jet lag can jeopardize athletic performance.

What makes these biological clocks tick? We know that these biological clocks are self-sustaining because rhythms of physiology and behavior persist in solitary confinement in many mammals, including humans[8]. One of the earliest studies in the field of circadian biology involved the recruitment of humans to live on a 28-hour day in Kentucky's Mammoth Cave. This study was done in 1938 but unfortunately for the field, only two research subjects were involved.

At any rate, the head researcher of the study—Dr. Nathaniel Kleitman who is regarded as the "father of sleep medicine"—found that certain physiological rhythms like body temperature continued to cycle every twenty-four hours even under a twenty-eight hour day. Seventy years after Kleitman's study, we have uncovered a significant amount of knowledge about these self-sustained rhythms of physiology and behavior, ranging from changes in mental performance, hormone release, and sleep quality under abnormal day schedules that are longer or shorter than twenty-four hours. As an undergraduate, I actually worked in one of these laboratories where we primarily studied how circadian rhythms of physiology change across puberty[9]. Working in this laboratory really made me appreciate the careful control that researchers take to keep experimental conditions constant.

However, in order to understand how these biological clocks tick, we have to return to the lab bench. Decades of work in animal models and thousands of cell cultures later, we know that biological clocks are self-sustaining due to a series of molecular events that shuttle genes out of the nucleus of the cell, turn them into proteins in the cytoplasm, and then interact with other molecular factors that inhibit further production. This is a classic example of a negative feedback loop, which essentially produces a self-sustaining cycle.

In recent years, we've been able to tag these ticking molecular factors to fluorescent markers to measure the rate and intensity of their ticking. This is called bioluminescence. Bioluminescence can be carried out in live mice or in dissected cells that can remain "ticking" in a dish if the proper chemical environment is maintained. Very recently, researchers at the Center for Skeletal Muscle Biology at the University of Kentucky have studied how the bioluminescent rhythms of the clock factor--Per2--which behaves as an inhibitory factor in this self-sustaining loop--changes in the muscles and lungs under an exercise program[10]. In this study, mice were presented with an opportunity to run on their wheels for 2 hours a day at a specific time for one month. Across this month, the time at which fluorescent Per2 cells were brightest, indicating that Per2 was highly expressed, progressively shifted towards the time of scheduled exercise. Of greatest interest, this exercise-driven shift only occurred in the muscle and lung, and not the "master" clock of the brain, showing how even in mice, there can be a disconnect between brain and body throughout the course of exercise.

What happens if you remove Per2 from the body? This was the basis of my dissertation in graduate school. The effect can be good and bad. I worked with mice wherein the *Per2* gene could be shuttled to the cytoplasm, but it could not be translated into a protein. Because of this, these mice were hyperactive. They would wake up two hours before the normal (wild-type) mice, and would go to sleep around the same as the normal mice. This well-structured sleep-wake schedule would give the Per2-mutant mice an extra two hours a day to run around. This makes them a great animal model of exercise, right?

However, instead of focusing on the good--exercise, we focused on the bad--addiction. The caveat of such hyperactivity is that they had extra opportunities to partake in other rewards, like alcohol[11]. I found that these mice were alcoholics, consuming the equivalent of a twelve pack every day with little intervention from the experimenter (me). The normal mice also consumed a decent amount of alcohol in a day--the equivalent of five beers—but the difference is that the Per2-mutant mice would binge drink; they would start drinking soon after the lights went off (they are nocturnal) and would continue drinking until the lights turned back on. They would also drink here and there during their (daytime) sleep period. Normal mice, on the other hand, only drank around the transitions of lights-off and lights-on.

While there is an explanation at the level of the brain for why Per2-mutant mice are alcoholics and are also more responsive to cocaine[12], which my lab mates and I found in later experiments, the levels of daily activity in these mice is nonetheless a contributing factor to why they drink more alcohol. If you're curious about the translational potential of this research, then you will be pleased to know that mutations of the Per2 gene in humans are linked to increased risks for alcoholism[13]. Whether or not these mutations also confer hyperactivity or gained athleticism in humans in unknown. I'd be inclined to start with the "fittest man on Earth"—Rich Froning—who brags about how he can't sit still for more than 15 minutes and also manages to do three workouts at significantly high intensity every day.

When to Train and Eat

Not everyone performs well at the same time of day. Our former educators know this, our bosses know this, and athletes certainly know this. The terms "lark" and "night owl" are constantly thrown around in the realm of health and fitness to refer to when people prefer to wake up, go to sleep, and be at their physical and mental bests. In the field of circadian rhythms, we call these chronotypes. Athletes are no different. We also have hard-wired chronotypes that are determined, in part, by the extent to which Per2 is expressed and when related biological networks are activated.

In fact, individuals with a distinctive mutation of Per2 have an increased risk for developing advanced sleep phase syndrome (ASPS). Individuals with ASPS are extreme "larks." They *prefer* to maintain a schedule similar to that of a "swing" shift worker, going to bed at dusk and waking up unassisted by an alarm clock in the middle of the night[14]. To the contrary, individuals with a distinctive mutation of Per1 are often extreme "night owls" suffering from delayed sleep phase syndrome (DSPS). In either case, athletes with a distinctive mutation of Per2 and Per1 are rarely presented with an opportunity to perform at their best, but I can guarantee that these individuals do respond well to 5:00 a.m. or 9:00 p.m. swim, track, or weight room training common of high school and collegiate teams.

Genes are not the only biological factors that determine when one should workout and perform at their best. Hormones released from various sources at the base of the brain, including the pituitary and pineal glands, also determine when we should perform well. As an undergraduate, I worked in a circadian rhythms, sleep, and performance-based laboratory for Dr. Mary Carskadon who is a pioneer in adolescent sleep research. Dr. Carskadon and her colleagues at Stanford and Brown universities discovered the reason why teenagers have proclivities for staying up past midnight and sleeping in past nine; puberty shifts, actually delays, the secretion of the hormone melatonin, which is referred to as the "hormone of darkness"[15].

Melatonin is the "hormone of darkness" because dim lighting or complete darkness initiates its release from the pineal gland in the brain. Have you ever felt abruptly tired in a dimly lit bar even if you were otherwise well rested? Melatonin was responsible. It is also available for over-the-counter consumption via pills and even chocolate brownies (The Original Lazy Cakes™). Melatonin is responsible for helping us fall asleep and stay asleep, aiming to minimize awakening across the night. In the sleep research community, we refer to this phenomenon as consolidation.

The problem with having a delayed release of melatonin driven by puberty, however, is that this perfectly natural biological process does not comply with school start times. As a result, this disconnect between biology and public policy can be detrimental to a teenager's physical and mental performances. In fact, school districts in Minnesota, Massachusetts, Washington, and Kentucky that have adopted later middle and high school start times have seen measurable improvements in scholastic performance as well as a reduction in car accidents en route to school[16]. But can a delay or inappropriate time of melatonin release also affect athletic performance?

Athletes understand that improving one's alertness, concentration, and reaction time are just as important as cardiovascular conditioning and weight training; there's a reason why highly caffeinated pre-workout drinks marketed towards improving alertness and concentration are widely consumed by athletes. I will consume them prior to training if I don't sleep well the night prior. Nevertheless, there is one study wherein researchers measured mental and physical attributes of athletic performance after a group of fit subjects consumed melatonin. Needless to say, the mental and physiological performances of subjects who were blindly given melatonin suffered; they had slower reaction times, poorer short-term memory, reduced alertness, and even a slower heartbeat during a four kilometer cycling task[17]. Therefore, I don't foresee melatonin being placed on athletic organizations' lists of banned substances.

While melatonin appears to somewhat sabatoge athletic performances, another hormone with timed release can enhance athletic performance: cortisol. Cortisol is released from the adrenal glands, but requires the brain in order to do so. This circuit is known as the hypothalamic-pituitary-adrenal (HPA) axis. It begins with the release of corticotrophin-releasing hormone (CRH) from a subdivision of the hypothalamus at the base of the brain. CRH then signals for the release of adrenocorticotropic hormone

(ACTH) from the pituitary gland, which finally allows cortisol to be released from the adrenals.

Cortisol is important for athletes because it stimulates the production of glucose in the liver, which is then transported to the blood and muscles. It also decreases inflammation. In fact, many athletes receive cortisone injections to reduce pain and swelling during their competition seasons. The actions of cortisol are often confused with another hormone released from the adrenals—adrenaline.

Unlike the "hormone of darkness," cortisol begins to rise when we wake up and continues to be pumped from the adrenals throughout the day. It's apparent that cortisol has a profound effect on athletic performance because a simple search for "cortisol and athletic performance" in a biomedical research database available to the public gave me 439 scientific studies of relevance with a majority of these studies examining the dynamics of cortisol release during competitive weightlifting, collegiate and professional strength and conditioning programs, or general psychological stress associated with competition. Much like melatonin, one should choose to train and perform around the time when our biological rhythms of cortisol are highest (and melatonin is the lowest). This is largely why our bodies don't often perform well in the early morning or late evening when cortisol release declines and melatonin release rises. Unless, of course, there is some travel across time zones which I will discuss later in this chapter.

Now that I've covered some genes and hormones and how they each affect athletic performance, it's time to discuss a biological process that is rarely neglected during the course of training: thermoregulation. The general physiological principle of thermoregulation is that when our core body temperatures (CBT) rises (i.e. we are hot), we sweat and when CBT decreases (i.e. we are cold), we shiver. Proper thermoregulation can benefit athletic performance; in the past year, a local business of Atlanta called Icebox Cryotherapy has sponsored my teammates and me. The company focuses on accelerating the recovery of athletes during a high volume of training. If you breakdown the word cryotherapy, it is exactly what it implies: recovery by extreme cold.

Basically, we stand naked in an upright tanning bed. For three minutes, gaseous nitrogen is expelled from a nozzle, causing our skin temperature to quickly drop. If you recall learning that the freezing point of nitrogen is wicked cold-- -340+ degrees Fahrenheit--from high school chemistry, then it's easy to imagine how quickly it took for us to shiver. The point of this rapid exposure to extreme cold for a short period of time is that it forces blood to be pumped into our muscles. Directing blood to the muscle is a limitation of ice baths—the traditional form of day-to-day recovery for athletes—because water can cross the skin, whereas nitrogen does not. The permeability of water forces blood to be shunted to the core instead of muscle so that core body temperature remains stable. Cryotherapy overcomes this biological drawback.

Cryotherapy has not only accelerated my recovery from high-volume training cycles, but it has also immensely improved my sleep. I'm not the most well rested morning riser, ironically enough, but cryotherapy certainly made me feel like one. To this end, we are currently in the process of designing a longitudinal study to examine the effects of cryotherapy on sleep quality in elite athletes.

Even without exercise and exposure to extreme temperatures, perfectly natural changes in our CBT affect how tired we are in the afternoon or at night. Our CBT waxes and wanes across the day and night due to the biological clocks of the brain. We have two dips in our daily rhythm of CBT with one occurring in the afternoon and the other in the early morning. Years of careful study have revealed that these dips in CBT increase sleepiness, providing biological justification (per your employer) for mid-afternoon napping. These dips are also associated with declines in mental and physical performance[18].

In laboratory and field settings such as US military bases, mental and physical performances related to afternoon and early morning sleepiness are tested by means of the psychomotor vigilance task (PVT). The exterior of the PVT resembles something that never left my side as a child--a Nintendo Gameboy. The machine provides precise measures of the ability to pay attention and records reaction times, which inevitably falter when one is sleepy. Subjects are asked to press a button as fast as possible and for as long as a light appears on a screen throughout several trials of testing. Any lapses in presses while the light is on or inadvertent presses when the light is off are measures of declines in performance. If you extrapolate these findings collected in the lab and at US government stations (NASA, Navy) to a ball or soccer field, performance will also likely suffer. Even though few studies have investigated the relationship between CBT, sleepiness, and athletic performance, lapses in attention and reaction time found in the laboratory and field are robust meaning that the results would likely translate to changes in athletic performance.

Sustained attention and infallible reaction times make for successful baseball players and soccer goalies who have random and often unpredictable bursts of activity for the few hours of practice or a game. But where in the brain is thermoregulation controlled? Creative anatomical studies undertaken in lower-order mammals such as rats and mice (relative to humans) have revealed that a subdivision of the hypothalamus—the medial preoptic area—is responsible. Thus, loss of the preoptic area unbalances thermoregulation as well as the ability to sleep[19]. The next step, of course, is to investigate how the medial preoptic area influences thermoregulation (and sleepiness) in humans. For now, we do know that yawning –a clear indication of fatigue--has some biological significance in that it may act to cool the brain and body[20].

What about eating? Westerners have dichotomous viewpoints about exercise and eating in that some people like me eat to train while others train to eat. Here, I'm only going to focus on the former group. Long before modern science, it's been known that the stomach controls hunger pains and appetite. In fact, there are stretch receptors lining the stomach wall that alert us to stop eating. This is one reason why vegetables help us lose weight; because four bunches of broccoli take up significantly more surface area and have 100-fold less calories than a package of pasta.

There are also two hormones—ghrelin secreted from the stomach and leptin secreted from fat cells--that tell us that we are hungry (ghrelin) or satiated (leptin). Leptin has been viewed as the next billion dollar weight loss "drug" in the biomedical research community because its secretion is directly proportional to the amount of fat one has[21]. Hypothetically, the more fat one carries around, the more leptin they secrete. This is a natural mechanism for preventing weight gain or maintaining physiological homeostasis.

However, weight management is not that simple in the modern world, which is why researchers are looking for other organs and biological substances that control hunger and appetite. There are several brain areas in the hypothalamus that are responsible for controlling hunger (ventromedial nucleus), satiety (lateral hypothalamus), and both (the arcuate nucleus). The arcuate nucleus is the team captain of this circuit by secreting large neurochemicals known as neuropeptides that are capable of binding to the lateral hypothalamus, ventromedial nucleus, and other subdivisions of the hypothalamus. This chain of events also activates the HPA axis—our body's stress system—to suppress appetite or inhibit the HPA axis to induce hunger.

The suppression of hunger by the HPA is fairly common for us; have you ever been ravished around lunchtime only to have your boss give you a pressing deadline? At some point during your state of panic, your hunger is forgotten.

Athletes most always experience a transient suppression of hunger during training and competition, especially if they are at an all-day swimming, track, or soccer tournament with double- and even triple-header matches. We also know that sleep suppresses our appetite. While there are a handful of people who make trips to the fridge in the middle of the night and at the very worst suffer from nocturnal eating disorder syndrome (NEDS)--sleepwalking coupled with "sleep eating" of high-calorie (and nasty) foods like butter sticks--hunger does not disrupt most people's sleep.

The mechanism of action for this is no different than what I have already described. From years of heavy training, I've also noticed that blocks of two- or sometimes three-a-day practices dramatically suppress my appetite. Although there is little literature as to why this occurs, I predict that the HPA is involved because a constant state of training puts the body in a constant state of physiological stress causing a constant need for repair and recovery. This is why very few people can handle heavy training year round without inevitably "burning out" and feeling like they've contracted mononucleosis for weeks on end (it sucks). Much like appetite, the brain also controls thirst. The signal of thirst originates in a special group of neurons in the pituitary gland known as organum vasculosum of lamina terminalis.

Much like hunger signals, thirst signals are also wired to the HPA axis. Athletes are well familiar with the experience of "dry mouth" or the constant need to drink water that come with pre-competition butterflies. As I mentioned in "The Athlete Brain in Competition," these pre-competition butterflies manifest from the HPA axis, which primes the body for intense exercise by mobilizing energy stores. Once the race or game starts, "dry mouth" usually disappears at least for anaerobic sports.

Sleep. You Need It.

Humans have evolved to be diurnal, the scientific term for being awake during daylight hours. We don't perform well at night. Terrible accidents in our society such as the oil spill from the Exxon Valdez and the nuclear spill of Chernobyl are textbook examples of this. However, I want to reserve further discussion on the timing of sleep and how it affects athletic performance for the latter half of this chapter when I talk about something most semi- and professional athletes are required to do, sometimes year round: travel across time zones. For now, let's talk about how much you should sleep and what type of sleep you should get. I have several family and friends who like to brag that they can function on four hours of sleep a night after which I remind them that "sleepiness makes you stupid"—a direct quote from a pioneer of sleep medicine, Dr. William C. Dement at Stanford University.

Actually, mouse models suggest that there could be a small (and lucky) population of people who only need four hours of sleep a night thanks to a mutation in the mammalian gene *Dec2* [22]. For the rest of us, the American Academy of Sleep Medicine and recent scientific studies recommend about eight hours of quality sleep a night. That means sleep with few disruptions from the noises of text message alerts and TV or a snoring spouse. In fact, a colleague of mine at the University of Pennsylvania has recently found that sleeping too much—more than nine hours a night—can be just as harmful for our physical and mental health as too little sleep[23].

The type of sleep one gets is also important because it directly impacts on how we physically and mentally perform. Mammals have two types of sleep across the night—non rapid eye movement (NREM) and rapid eye movement (REM) sleep. NREM sleep dominates the first half of the night. NREM sleep is usually deep and restorative, meaning that it can be difficult to wake someone up from this type of sleep; children are the best example. NREM sleep is also where tissue repair and growth occurs after a day's work.

This is because of the secretion of growth hormone from the pituitary gland. Children secrete more growth hormone than adults, but no one can deny that athletes need it, particularly professional doping committees. The number of professional, collegiate, and even high school athletes who have taken growth hormone supplements, producing abnormal levels, is significant with baseball and football players being the most common and repeat offenders. With the amount of weekly travel coupled with the nature of training that athletes beat themselves up over for decades, doping by means of growth hormone is not shocking. Of course, this isn't the rationale for why athletes dope with growth hormone—rather to quickly re-grow and repair tissue--but I do wonder if this type of rationale supported with appropriate scientific evidence would hold in court?

We will save this discussion for the last chapter of the book. At any rate, we can agree that NREM sleep and the subsequent release of growth hormone is important for physical performance. Brain control of NREM sleep is diffuse. Areas of the hypothalamus, cortex, and brainstem work together to release neurochemicals such as serotonin, GABA, norepinephrine, and histamine. Recent studies have also shown that the phenomenon of NREM sleep is not always global. In fact, when we are tired after a night of poor sleep or from the dip in our CBT in the afternoon, individual neurons in these brain areas can go into a deep sleep ("offline") at random intervals. For a baseball player or soccer goalie, this may be the difference between a game won or lost.

In contrast to NREM sleep, REM sleep dominates the second half of the night. REM sleep gets its name from the spontaneous, synchronous eye movements that occur during this stage of sleep. REM sleep also initiates muscle paralysis at the level of the spinal cord except for the muscles of the inner ear (for waking up in the face of danger), eyes (for synchronous eye movements), and diaphragm (for breathing). While the deactivation of most muscles of the body can be restorative in some sense, the biological significance of REM sleep lies in its ability to consolidate factual, emotional, and motor memories. It's also the *only* stage of sleep where individuals dream.

When people are deprived of REM sleep, their mental (e.g. recall of words and faces) and motor (e.g. proficiency at finger-tapping and video games) skills deteriorate. Much like the PVT, which aims to study the relationship between sleepiness and performance, studies of REM sleep and its necessity for controlling motor memories is limited to the laboratory and has yet to be studied in athletes. I learned about one study of such sorts at our annual sleep research conference a few years ago. In this study, researchers at Harvard asked subjects to play a first-person ski game much like you find at Dave&Buster's[24]. Why was this done?

Peak performance in first-person shooter or sports games is proportional to the amount of practice and memory encoding that occurs at the level of the brain and muscles. The brain learns to predict when a zombie will pop out or when a sharp bend in a race track appears and muscles learn to predict when and how often to fire a gun or how much to lean on a bike. After a few practice sessions, the subjects in this Harvard study were asked to re-try the game after taking a nap; napping is uncommon in Western culture, but when people do nap mid-day, they often find themselves in REM sleep, even if the nap is only twenty minutes in duration. The reason for this is because the dip in CBT in the afternoon favors REM sleep. I actually studied this relationship between CBT and propensity for REM sleep in the circadian rhythms, sleep, and performance-based laboratory for a senior research project. As for the Harvard study, subjects who took a nap improved their virtual athletic performance while those deprived of a nap did not. Similar studies of motor memory and changes in motor skills have also been done in overnight studies wherein subjects are deprived of REM sleep at night.

If we extrapolate these laboratory findings to athletes, the benefits of napping are clear. Not only can a brief period of sleep allow the athlete to re-fuel before their next practice, but napping can help encode the skills they learned in the most recent practice, making for better footwork or a new personal best in the weight room. Much like NREM sleep, REM sleep is controlled by specific brain areas. The defining electrical activity or EEG of REM sleep is often described as PGO waves. "PGO" references areas of the brain where the EEG signal originates—the pons (P) at the base of the brainstem, the geniculate (G) region of the thalamus in the middle of the brain, and the occipital cortex (O) on the back of the brain.

The pons, geniculate, and occipital cortex work together to transition in and out of REM sleep and rely primarily on the neurochemical acetylcholine to do so. In humans, scientists in Europe have been able to manipulate the amount of time we spend in REM sleep through pharmaceuticals that enhance or inhibit acetylcholine communication[25]. A side effect of doing this is that it also manipulated performance on a motor task: the speed and accuracy at which the subjects typed a sequence of numbers on a computer with one hand. This type of manipulation could definitely be categorized as next-generation doping, extending beyond the improvement of motor skills with magnetic stimulation of the brain as I discussed in "The Athlete Brain."

But do we need both NREM and REM sleep to perform well? Yes, even if the sport of choice is not as physically demanding or debilitating, requiring constant tissue recovery and repair. I mention this because there has been increased interest for the science-based practice of polyphasic sleep. This was popularized by Timothy Ferriss in his internationally best-selling book and blog the *The 4-Hour Body*. The rationale behind polyphasic sleep is that you have extra time to do other things in a day by restricting yourself to two hours of REM sleep spaced out in six, 20 minute intervals—the daily average for most humans. These two hours of REM sleep serve as a means for your body to encode a day's worth of memories and learned skills, but allows you to go without the other 6 hours of NREM sleep which Ferriss and his scientific colleagues deem as unnecessary or a waste of time.

Polyphasic sleep appears to be justified because it requires an understanding and exploitation of the predictable daily rhythm of REM sleep driven by dips in CBT. The science behind polyphasic sleep is very attractive at first. Who wouldn't want to have more time in a day for research and training without a compromise in brain function? But nearly every biological process serves a purpose, including NREM sleep; NREM sleep provides an opportunity for your nerve cells "to rest" at random intervals and your muscles to replenish energy stores. In fact, Timothy Ferriss provides a disclaimer about the dangers of deviating from a polyphasic schedule, and how it can screw your body over to the point that you end up "wasting" time in the end; by getting terribly sick or suffering from chronic fatigue. In the end, I don't foresee many athletes jumping on the polyphasic sleep bandwagon for the purpose of having more hours in a day to train because it's selective to your brain and not muscles. Even the "dumbest jocks" can recognize this.

Now that I have outlined the importance of sleep for athletic performance, what happens to athletes who are deprived of it? Fortunately, we have many public databases with this information. Many of these studies were conducted at Stanford University which not only has an elite group of researchers in sleep medicine--the field originated at Stanford and University of Chicago--but Stanford also boasts an elite group of student-athletes and alumni: Tiger Woods and Andrew Luck to name a few. Overall, these studies have found that the extent of insufficient sleep directly correlates with declines in scholastic and athletic performances. This is not a surprise. I did happen to find one study though in which the "athletes" performed better under conditions of short-term sleep deprivation; unfortunately, they were race horses[26].

Horses aside, insufficient sleep on the night before a competition is normal. I can only think of a few instances where I have had restful sleep uninterrupted by pre-competition butterflies the night before conference, districts, regionals, state, national, and international competitions. I've been a seasoned athlete for over twenty years and poor sleep on the night before competition happens without fail. My most recent bout of pre-sleep competition butterflies was also the most unbearable because the competition was three days long: the Reebok Crossfit Southeast Regionals and then Crossfit Games. Every night, I would obsessively envision myself going through the next few workouts with little room for error.

Professional athletes have reported the same type of pre-competition OCD-like meditation; the three-peat "Fittest Man on Earth," Rich Froning Jr, has talked about the little sleep he gets during the three days of the Reebok Crossfit Games as well. Needless to say, most of the studies are on-campus, meaning the athlete is asked to fill out the questionnaire or is outfitted with EEG signals for an overnight in the lab after a night of practice, not competition. To conclude, I'm not stating that a night of insufficient sleep (versus rested sleep) has no detrimental impact on performance (because it probably does), I'm just emphasizing that you'd be hard-pressed to find an athlete—amateur, seasoned, or professional--with rested sleep the night before a major competition. It's part of the game.

Game Day Travel and the Prescription for Better Performance

Since 2008, I have consulted with Olympic teams prepping for travel to Beijing and London. Why? The body is cantankerous about adjusting to a new light-dark cycle, especially one that is more than three hours ahead or behind. You don't need scientific studies to realize this. You can simply look at the revenues of pharmaceutical companies who sell drugs marketed towards shift workers and frequent flyers that improve sleep (Ambien) or increase alertness (Nuvigil) and annual over-the-counter sales of melatonin. But science does help us understand what happens to our brain when we enter a new time zone. There are plenty of clinical and epidemiological studies that have examined the detrimental effects of shift work and trans-time zone travel on health ranging from increased rates of cancer, mood disorders, addiction, cardiovascular trauma, and unhealthy weight gain with the inability to sleep being one of the more immediate, common complaints.

As a freshman in college, I competed in my first international competition against track and field programs in the greater London area. This was also my first time in Europe and my first time in a time zone more than an hour apart. The immediate effects on my training were evident. Photo by Allison Brager.

As I mentioned earlier, all mammals have NREM and REM sleep and transition into NREM and then REM sleep in a similar manner. The exception would be aquatic mammals such as dolphins where one half of their brain sleeps while the other half is awake; this is called unihemispheric sleep. So, to study sleep in mice exposed to routine jet lag is not unreasonable. My colleagues and I completed a study of such sorts. In this study, the mice were flown to Paris wherein their night was advanced by six hours every week for three months. When we looked at day-to-day changes in sleep, we found a 40 percent loss of daily sleep within three days of the shift in the light-dark schedule[27]. Further, the greatest loss in sleep was not on the first day, but rather the third day.

We also found that this trip to Paris activated neurons in brain centers that control sleep-wake in a similar manner to animals that had been outright sleep-deprived for 6 hours. This information is why Olympic and professional teams have recruited sleep researchers in preparation for game-day travel. To give you an idea, scientists have worked alongside Olympic committees to develop sleep-friendly rooms for the Olympic village that are not only sound-proof, but also light-proof since light is the most potent disruptor of circadian rhythms in mice and men. Even turning on the bathroom light at night is enough to shift the body full of clocks that I talked about early in the chapter.

But let's transition from studies in mice to studies in men because there actually is a handful of information on jetlag, sleep deprivation, and their consequences on athletic performance. Most of the laboratory-based studies have focused on shifts in rhythms of melatonin and CBT, which are easy to measure with little invasiveness.

As I discussed earlier in the chapter, the relationship between CBT, sleepiness, and mental and physical performance is intimate; drops in CBT increase sleepiness which lead to declines in mental and physical performances. With a new light schedule, especially one that is more than a few hours ahead or behind, the release of melatonin—the "hormone of darkness" that is sensitive to light— is suppressed and the normal afternoon and early morning dips in CBT occur earlier or later.

Obviously, both of these physiological shifts wreak havoc on sleep and performance by keeping people up at night, waking people up too early, and causing a lack of focus and energy throughout the day. In fact, one study has compiled a breakdown of motor and mental skills that suffer from insufficient sleep driven by game day travel; for low-aerobic sports that require high levels of alertness and fine motor skills like sailing and archery, there's more room for error. For team sports that require high levels of concentration, there's poorer decision-making. For individual sports with a mixture of aerobic and anaerobic movements like swimming, mixed martial arts, and weightlifting, there's a loss of power and quicker time to fatigue. Although these symptoms and changes in our body's physiology from game day travel across time zones eventually disappear, the problem is that it takes at least a few days for the body to adapt.

For professional baseball and basketball players, this is an unresolvable problem because many teams are already on their way to a new time zone to play another team. As an example, the 2012 NBA season, which was truncated by contract agreements, had more game-day travel than in year's past to make up for lost inter-conference playing time. The season was also plagued by poor refereeing, increased amounts of injuries, and complaints of constant exhaustion. Although no empirical studies were undertaken, sports pundits and researchers like me believe that jet lag and constant weekday and weekend game travel were the culprits[28].

In the field of sleep research, we often talk about "sleep debt" which refers to any negative deviation from your body's desired amount of nightly sleep. So if your body requires eight hours and twenty minutes of sleep a night (the human average), but you only get seven hours, you will need to make up that 1 hour and 20 minutes at some point. If we short-change ourselves 1-2 hours of sleep for a few days once a year, this isn't a big deal. We may be sleepy for a few days, but we will rebound quickly. But if we are depriving ourselves of 1-2 hours of sleep a few times a week for months or years, we aren't getting quality sleep.

This is largely because our body full of clocks keeps getting exposed to different light-dark cycles, which is a serious problem. Under these circumstances of chronic sleep deprivation, our sleep debt creeps up on us whenever we aren't socially stimulated such as when we are driving, watching a movie, or standing around in the outfield. Even if we aren't sleepy *per se*, our minds are slower due to individual neurons going offline at random intervals to "rest," and our physiological rhythms of growth hormone which acts to repair damaged and maintain healthy tissue can be suppressed. This is one of many reasons why game travel in collegiate and professional sports communities should be re-evaluated or more carefully planned with the help of researchers in sleep and circadian rhythms.

Does the length and direction in which we travel also make a difference? That is, does performance suffer worse upon travel to the Pacific versus Mountain time zone from the East Coast, or do West Coast teams playing a night game at home have a physiological advantage over their East Coast opponents or vice versa? If you look at the regional distribution of professional NFL, NBA, and MLB teams across the US, nearly half the teams lie in the Eastern US time zone (Atlantic coast) while a majority of the others are in the Central or Pacific US time zones (Midwest or Pacific coast).

Within a professional organization, there are two conferences that are not solely determined by region of the country; both the MLB and NFL are divided into American and National Leagues which each have teams on the West Coast (Oakland A's for American and San Diego Padres for National) and East Coast (Boston Red Sox for American and New York Mets for National). This is particularly bothersome for Major League baseball players who play 2-4 games per week, sometimes in different time zones, versus a professional football player with one game on a

Monday, Thursday, or Sunday. Luckily for us, researchers have investigated if the type of game (away versus home) and direction of travel (east versus west) impacts athletic and mental performance through careful examination of win-loss records, changes in individual statistics, and even a few questionnaires on sleep habits.

Even with mice and hamsters, the direction of travel makes a difference; mice have an easier time adjusting to delays in their light-dark cycle, whereas hamsters have an easier time adjusting to advances. This time zone preference actually isn't random either, but rather explained by how closely biological clocks "tick" towards twenty-four hours. Because the biological clock of a mouse "ticks" slightly less than twenty-four hours, a delay in an environmental schedule would make their clocks "tick" at twenty-four hours. Meanwhile, because the biological clock of a hamster "ticks" slightly more than twenty-four hours, an advance in an environmental schedule would make their clocks "tick" at twenty-four hours. Human biological clocks are similar to those of a hamster in that they "tick" slightly longer than twenty-four hours.

This is why most humans have an easier time with Western versus Eastern travel. Athletes are no different. One of the first studies to investigate athletic performance following East versus West Coast travel was determined from archived data of the 1996 collegiate (NCAA) football season; the University of Florida Gators were national champions that year. Overall, teams that traveled more than one time zone eastward performed consistently worse than teams traveling more than one hour westward: they scored fewer points, allowed more points, and had greater point spreads even when controlling for progress of the game: 1st quarter versus 4th quarter[29].

In terms of whether travel to the Pacific coast versus Midwest for a New York Yankee matters, it does. In 2008, I learned about such a concern at our annual sleep conference held in Baltimore, Maryland through Dr. W. Christopher Winter at the University of Virginia who happened to have a research poster next to me. Dr. Winter and colleagues entered the scores of over 24,121 MLB games into a database that controlled for number of time zones travelled. Independent of direction, 60 percent of games were lost if the baseball player traveled three zones and 52 percent were lost for travel through two time zones. An 8 percent difference may not seem significant, but in the world of professional sports that can be the difference between finishing last in the league (and subsequently getting first pick in the next year's draft) and vying for a pennant. These results--that athletic and mental performance suffer more as the number of time zones travelled increases--make biological sense because that extra hour of light (or dark) for one more time zone would mean that your body would need an extra day and sometimes more to adjust. However, there are many strategies for adapting quicker to a new time zone of which I'll discuss next.

As for whether certain coastal teams have an advantage during night games--the games with the highest viewer ratings--this also matters. The sleep community of Stanford

University conducted a carefully controlled examination of win-loss records across forty years of play in the NFL to determine if East Coast teams had a biological advantage over West Coast teams during a night game played on the East Coast or if West Coast teams had a biological advantage over East Coast teams during a night game played on the West Coast.

Unlike previous studies, these researchers also factored in the Vegas point-spread which is carefully calculated: it is based on win-loss record, injury reports, historical matchups between the teams, weather, and whether it is an away versus home game plus many other variable. Therefore, a "win" in this Stanford study was defined as winning the actual game *and* beating the Vegas point-spread. Unfortunately, East Coast teams are always at a disadvantaged when playing a night game with West Coast teams winning 70 out of the 106 night games played regardless of whether it was an away or home game. So even if an athlete is at a biological disadvantage to performing at their best due to game day location, can modern science and medicine help?

Millions of Americans complain about not getting enough sleep or quality sleep, but few people possess the knowledge or are willing to put forth the time and effort to remedy their sleep problems. Athletes are no different. Even someone like me, who possesses far more knowledge about the subject, can be a hypocrite and treat a sleepless night as a pre-competition ritual. But there are ways to ensure that you do get a decent night of sleep prior to a competition if your game day travel is favorable and allows you several days to adjust to a new time zone. Most of these strategies begin weeks prior to a competition, but in the process, they will also readily enhance your training potential.

The easiest and most motivating strategy is to keep a sleep diary, which simply documents how much you sleep, the time you went to sleep, and how you felt in the morning. I didn't realize the power of a sleep diary until I took a psychology course on sleep under the tutelage of my undergraduate advisor as a junior in college. We had to keep a sleep diary throughout the semester. Every week, Dr. Carskadon would present the class average of daily sleep amounts on the weekdays and weekends which started off promising, but of course waned as the end of the semester approached. Whether it was out of academic and group pressure to maintain a sleep diary or a personal interest in sleep, I quickly began to strive for the golden 8.4 hours of sleep a night. I would also wager that it is no coincidence that I had a successful indoor and outdoor track season my junior year.

Another non-clinical strategy for dealing with game day travel, particularly across time zones, is to gradually delay or advance your normal bedtime schedule across a week of competition depending on the direction of travel for the competition. As I discussed earlier in this chapter, several environmental and social stimuli such as light, temperature, and traffic help to cue the body and brain when to sleep and wake up. Therefore, by gradually misaligning your routine bedtime schedule with these environmental and social stimuli prior to a competition, the actual shift for game day travel is likely to be less of a shock to your body and brain. This is a strategy that has been recommended to Olympic athletes traveling across continents. Lastly, power naps are recommended. A 20-40 minute nap in the afternoon can boost alertness, mood, and even hand-eye coordination as studies in college students and even student-athletes have shown. Plus, a power nap in the middle of the day isn't enough to shift your biological clock and entirely disrupt your sleep that night.

Within the clinical world, there are two strategies that are recommended to travelers and shift workers. Both can be inexpensive pending that you have medical insurance, but both require routine feedback from sleep consultants and physicians. Sleep medications are the first but less recommended option. There are a variety of drugs that have been manufactured to ameliorate jet lag. Many act to increase alertness while others act to have a deeper night of sleep. Modafinil falls into the former category and increases alertness by means of increasing the amount of dopamine: the neurotransmitter linked to the experience of excitement, pain, and pleasure of which I also have tattooed on my forearm.

Unfortunately for athletes, modafinil appears on several banned substances lists. For deeper sleep, benzodiazepines have been the golden drug for years until patients started complaining of severe and long-lasting episodes of sleepiness the next morning and afternoon. Since the early 2000s, non-benzodiazepines have entered the drug market, but their side effects are also scary; every year at the annual sleep meeting, I constantly walk past posters that document incidences of people walking, online shopping and gambling, and even having sex and driving while asleep because these drugs seem to activate brain pathways that regulate habitual behaviors. Many of these sleep aids act through GABA: the classic inhibitory neurotransmitter found in many pockets of the brain. It was first discovered in the lobster. The bottom line is that pharmaceutical drugs aren't the best approach for improving sleep for the benefit of athletic performance.

However, there is at least one natural, pharmacological aid that has been taken by athletes to improve sleep and athletic performance. I first encountered this sleep aid in the exhibit hall at the annual sleep meeting: tart cherry juice. Tart cherry juice acts as an antioxidant aimed to remove cancer-linked free radicals that we are exposed to as part of modern life. It also contains high levels of melatonin. Two studies conducted in the UK and US have shown that tart cherry juice improves the quality of sleep in healthy and clinical insomniac populations[30,31]. In the athletic community, tart cherry juice has also gained recognition for increasing one's endurance capacity and reducing muscle pain (note: Trader Joe's has the most affordable tart cherry juice).

The second strategy that clinicians recommend for adjusting to game day travel across time zones and subsequently improving sleep quality is blue light exposure. As I have discussed throughout this chapter, light is the most potent stimuli capable of shifting our biological rhythms. Our bodies are ultra-sensitive to "cool" light, which have short wavelengths and marginally sensitive to "warm" light which have long wavelengths. Because blue is a cool color, blue light jolts the biological clock, causing our physiological and behavioral rhythms to rapidly shift. Again, this is a reason why you should refrain from turning on the bathroom light when nature calls in the middle of the night.

In humans, the general idea is that 10-15 minutes of bright blue or white light in the morning can advance the time we go to sleep the upcoming night and 10-15 minutes of exposure in the evening can delay the time we go to sleep that night by re-organizing the physiological pathways that regulate sleep. These sleep-improving effects of blue and white light have been consistently shown across years of clinical and basic sleep research. For athletes, I advise that you use this strategy with caution in the absence of a sleep consultant. At the other end of the spectrum, exposure to "warm" light like red light at night has been shown to improve sleep quality in professional female basketball players in China[32]. Whether or not there was a complementary improvement in their athletic performance is unknown.

As I conclude this chapter, I hope that you can realize the importance of habitual sleeping, eating, and training for better athletic performance; the time at which you chose to sleep is just as important, if not more, than getting adequate sleep. During the week of game day travel, adequate sleep is even more important. But if life prevents you from getting appropriately timed or more sleep leading up to a competition, you can always rely on new discoveries in science such as bright light therapy and healthy foods like tart cherries to mask this insufficient sleep. Lastly, don't underestimate the power of a daytime nap. Many companies are incorporating this into the daily work routine for a very good reason. It boosts mood, the ability to recall factual information, and fine-tunes motor movements. Sleep on it.

Chapter 5

Central Doping

In recent years, it seems as though every professional athlete is a skilled cheater, exploiting the science of exercise physiology and pharmacology and hiding the circulation of substances from doping officials. Each sport seems to have a common modality for doping, ranging from steroid and growth hormone abuse in baseball, the extraction, storage, and re-infusion of red blood cells in cycling, to the use of some peculiar deer antler spray in American football. Although all of these substances are meant to affect the muscles and metabolic enzymes, what effect do they have on the brain? Of greater interest, how are substances and techniques that are specifically designed to act on the brain enhance athletic performance? As a disclaimer, I am not advocating a means for athletes to cheat the system by taking substances known to work for scientists, but also listed as banned by doping agencies although I will discuss a few of these performance-enhancing drugs (PEDs). Instead I'm advocating for athletes to take advantage of science that has already received prior approval for use from doping agencies. I do not think there is anything wrong with using science to get ahead as long as it is done with integrity and safely. This chapter will explore these brain boosters.

Daily Java

Caffeine is the most widely consumed drug in the world. Over 2.25 billion cups of coffee are served a day around the world. This datum does not even include other caffeinated beverages like energy drinks that have dominated the beverage market in the past few years. Adults aren't the only ones consuming caffeine. It has become the drug of choice among adolescents, but not without parental and public health concerns. I was first introduced to the outrageous idea of kids and teens drinking caffeine, in particular coffee, upon moving to the tiniest state in the Union—Rhode Island—at the age of eighteen. After spending a few weeks in Providence, I came to learn that coffee milk or milk infused with coffee-flavored syrup was the official state drink. It was freely dispensed at my college's dining hall and bottles of coffee syrup were widely sold in grocery stores. It is also near impossible to not find a Dunkin' Donuts more than two miles from any house in the state. Although it took me a while in college to acquire a taste for coffee, I began to drink it in college because of its mental and physical performance-enhancing benefits.

Caffeine is a stimulant meaning that it acts to excite rather than depress nerve function in the brain. It makes sense that academia, medicine, and other professions requiring high levels of alertness and communication are among the most frequent coffee consumers[1]. In the brain, caffeine blocks the actions of a specific signaling cascade: adenosine. Adenosine is produced and released from glia, which make up 90 percent of the human brain because glia provide structural support, physical protection, and nutritional sustenance for nerve cells, and help to remove pathogens.

Adenosine is largely responsible for sleepiness, at least physiological sleepiness, wherein adenosine causes nerve cells that are responsible for keeping us awake to fire less and nerve cells that are responsible for sleep to fire more. Caffeine, however, is capable of blocking adenosine-induced sleepiness, allowing nerve cells that drive wakefulness to fire longer. Caffeine also has a long half-life; a standard measure of drug action in the field of pharmacology that is used to characterize how long a drug acts in the brain and body. Caffeine's half-life is five hours, which means that a cup of coffee can act on the brain for roughly ten hours.

A recent study that has grabbed the attention of sleep research and Crossfit communities found that the time at which one consumes caffeine and works out can have a significant impact on power and stamina, particularly for weighlifters[2]. These exercise physiologists recruited subjects with experience in power lifting: squatters and benchers. The researchers compared power output generated by major muscle groups while the subjects back squatted or bench-pressed 75 percent of their best weight. This was done either in the morning at 9:00 a.m. or in the evening at 7:30

p.m. During this time, the subjects took a capsule containing a high amount of caffeine--3 mg per kg of body weight--equivalent to two cups of strong coffee. Individuals training in the morning had lower power output compared with those lifting in the evening. However, someone who took a caffeine pill in the morning could improve performance to the level of someone lifting in the evening without caffeine.

The researchers also examined changes in sympathetic nervous system, which aims to provide constant and rapid supplies of glucose derived from the liver to the muscle and brain. There are many physiological identifiers of change in the sympathetic nervous system, including norepinephrine, epinephrine, cortisol, and testosterone. All of these biochemical markers were measured in this study. As expected, lifting significantly increased activity of the sympathetic nervous system from levels found prior to the workout. Much like the results of muscle performance, subjects lifting in the morning had less activation of their sympathetic nervous system relative to their evening counterparts unless they were given caffeine To recapitulate, this study suggests that a strength-biased athlete is likely to see the largest gains in pound for pound strength if they train in the evenings or after a pre-workout shake in the morning, which have a significant amount of caffeine.

The caveat of this study, at least from the perspective of sleep researchers, is that one's chronotype or time at which an individual naturally performs at their best relative to when they go to sleep and wake up was not considered. In the last chapter, I discussed a group of individuals who have unusual bed and wake times due to genetic mutations. If it just happened that one of these participants suffered from advanced sleep phase syndrome wherein their evening is most people's morning, then the results could be skewed. However, I trust the scientific enterprise and believe this was not an issue for the study.

Despite the benefits for muscle and mental performance derived from caffeine, people develop tolerance to caffeine extremely quickly and also suffer from caffeine withdrawal extremely quickly. We have all met someone who is highly volatile and suffers from a headache if he or she doesn't have a cup of coffee by lunchtime. I am no different. The biological reason for headaches caused by a lack of daily caffeine is fairly straightforward. Caffeine causes blood vessels of the brain to constrict—which is why consuming caffeine immediately prior to a doctor's visit may overestimate your true measurement of blood pressure. In the absence of caffeine, the blood vessels overcompensate and dilate too much thereby causing a headache. Sleepiness is also amplified in the face of caffeine tolerance or withdrawal.

This is likely due to the overcompensation of the adenosine system. This concept has never been directly investigated in humans, but a change in the adenosine system with caffeine tolerance and withdrawal and its effects on sleep has been investigated in cats[3]. At any rate, the consumption of caffeinated beverages is prevalent in many athletic communities. It is most beneficial for sports requiring high levels of focus and concentration despite low levels of physical activity like America's favorite pastime: baseball.

So how much caffeine do you need to take in order for it to enhance athletic performance? In the world of exercise physiology, caffeine is widely known to increase endurance by means of initiating the breakdown of fat, which is a long-lasting fuel source. In the nervous system, caffeine decreases the perception of exertion by means of making it easier for the neurons innervating the muscles to be activated[4]. Many studies show that doses of caffeine equivalent to a cup of coffee (100-150 mg) can have a performance-enhancing effect. The caveat is that tolerance destroys this performance-enhancing effect. This means that a person who drinks a cup of coffee a day is unlikely to experience a boost in endurance from that one cup. They would have to drink two or more and so forth. This strategy may seem reasonable, but there are cardiovascular, digestive, and other health risks related to the overconsumption of caffeine.

As for athletes, many drug-doping agencies place strict limits on caffeine consumption during a competition or throughout the sporting season. The limit for the NCAA is 500 mg at the time of drug testing, which is about five cups of coffee. While it seems absurd that anyone would consume that much coffee prior to a competition, there are many pre-competition workout drinks aimed to improve focus, power, and exertion wherein a single serving is far more than 500 mg of caffeine. This is ludicrous, and even more ludicrous that the Food and Drug Administration does not regulate health and PED supplements unlike most food products.

Because of this lack of regulation, companies selling health supplements are not required to put what their products contain or the dosages of the active ingredients on their labels. Most importantly, a lack of regulation coupled with improper knowledge of what is being consumed and what the ingredients are doing have caused tons of athletes to fail drug tests, banning themselves from the competition or worse, future play. For professional athletes, this has a direct effect on income. I have seen this happen in collegiate and Crossfit communities on *numerous* occasions with overconsumption of caffeine being a primary reason for why an individual or team are penalized, albeit short- or long-term, for consuming an otherwise "safe" beverage. I imagine that failing a drug test due to overconsumption of caffeine will continue to be on the rise since the beverage market is saturated with all sorts of highly caffeinated products, including coconut water. For athletes, be mindful of what you are exactly taking, the amount allowable, and seek out any one with expertise in chemistry or pharmacology.

Age of Taurine

In 1987, a unique type of beverage was created. Nearly twenty-five years later, Redbull has become one of the most widely recognized and consumed carbonated beverages on the road, in bars, on college campuses, and even athletic competitions. The company sponsors several athletes in a variety of athletic circles, but primarily caters to the extreme sports athlete. Sales reps for Redbull have hung around the athletic complex of my undergraduate university from time to time. I quickly realized that these Redbull reps would have been better served on the campus quad down the street because a carbonated beverage prior to an endurance-based competition does not often agree with one's digestive system even if there could be an endurance-promoting effect. Aside from caffeine, the other important ingredient of Redbull and related energy drinks is taurine. Taurine is not just a supplement. It is a biologically active compound widely made throughout the body. It is very important for the makeup and function of the nervous system as well as muscles.

At the level of the brain, taurine is vital. It balances the amount of electrolytes, coordinates the firing of large groups of neurons, and helps to prevent premature death of neurons due to over-excitation: a condition known as neurotoxicity. Thus, it's no wonder why taurine is blended with caffeine for the purpose of enhancing athletic performance, at least mentally.

Physically, taurine improves endurance; it keeps the skeletal muscles well oiled by increasing the clearance of lactic acid, which accumulates in the muscle over time with intense exercise, and elevates the availability of ATP for the muscle, which you may recall as the "powerhouse of the cell"[5]. What this means is that taurine reduces time to fatigue and helps to maintain power output.

Mentally, taurine also reduces the perception of exertion, allowing an athlete to accomplish a higher volume of training.

Despite its brain-and muscle-fueling effects, taurine is not considered to be a banned substance by drug doping agencies. Perhaps it is because the impact of taurine on athletic performance is not conclusive or perhaps the effects are marginal. At any rate, blending taurine with a reasonable (and legal) amount of caffeine may be one means for athletes and their brains to get a quick boost in performance during competition.

Suns Out, Guns Out

Whenever a drug doping scandal is brought to the public's attention in the sport of baseball or football, one of two drugs is the common the source of cheating: growth hormone or anabolic steroids. Both can have tremendous impact on brain and bodily functions although through very different mechanisms. One difference related to reasons for use is that growth hormone is more prominently used to accelerate recovery whereas anabolic steroids are used to become stronger and bigger during the off-season. In my own research community of sleep, growth hormone is an attractive biological factor. When we are young, we secrete a lot of it. Do you ever wonder why it takes an eight-year old a day to recover from a deep bone or muscle bruise, while adults hobble around for weeks?

Growth hormone is released from the pituitary gland—a testes-like structure hanging from the bottom of the brain just above the roof of the mouth. Sleep is one cue for the release of growth hormone, but not just any sleep. Growth hormone is primarily released during deep non-rapid eye movement sleep (NREM) found during the first half of the night wherein the nerve cells of the brain harmoniously fire in tandem creating big, swooping waves that sleep researchers like myself can smile at. In the research community, we refer to this form of deep sleep as slow wave sleep (SWS).

The release of growth hormone during SWS is one of many reasons why it is important to not skimp on NREM sleep. This is especially true for children who need growth hormone in order to grow properly. In fact, there is one study in the field of endocrinology—the biology of hormones—that found that misappropriate release of growth hormone during sleep at night was linked to short stature in children[6].

It also makes sense that anyone who goes through a period of training and competition wherein rapid tissue repair is needed would have an increase in SWS that directly corresponds with the intensity of exercise[7]; bodies in training need to recover and repair as quickly as possible. Fortunately, growth hormone allows this to happen. Once this growth hormone is released from the pituitary, it acts on many of the bodies' tissues to restore and repair, not just muscle and bones, and helps to increase muscle mass and bone density for as long as training and competition occur.

Although growth hormone has been widely abused among baseball and football players because of its anabolic (muscle- and tissue-building) effects, is there a way for athletes to increase internal production of growth hormone without resorting to banned substances? One strategy that would not require the intake of supplements or pharmaceuticals is to optimize the amount of NREM sleep you get. As I discussed in the last chapter, the type of sleep one gets is dependent on two, major factors—the environment and physiology. For most collegiate athletes, game day travel is the day of the competition, which is usually a Saturday. On the night before a game, collegiate athletes not only have to adapt to going to bed earlier and waking up earlier than they even would for a class during the week, but they also have to deal with the weekend party noise of their non-athletic dorm or house mates.

These three factors—a change in one's normal sleep schedule, ill-timed exposure to light in the morning, and incessant background noise—can wreak havoc on the sleep cycle throughout the night, but also across the next week. Since most teams compete and travel for 6-8 weeks on end, some athletes are likely to never adapt and excel as a student-athlete, or get the restorative sleep that they need. That being said, how can an athlete optimize their restorative, NREM sleep and more importantly, optimize the release of growth hormone?

The most obvious answer is to keep a habitual sleep schedule, and a sleep schedule that agrees with our biological makeup of being a diurnal (day-active) species and not the conveniences of the modern world. This means going to bed at the same time every night, particularly in the late evening and waking up at the same time every day, preferably when the sun rises. There are plenty of epidemiological studies—which require a huge sample population of at least 1,000 subjects—pointing to the contribution of unusual bed times on sleep quality. Part of the reason is because our biological clock deep in the hypothalamus also drives the quality of our sleep.

When the biological clock is on high alert cueing us to wake up, we aren't going to enter or stay in the deeper, restorative stages of NREM sleep. So for those of you who go to bed in the wee hours of the morning on a "Saturday" night, chances are you didn't get as much of the deep, restorative NREM sleep as you would on a weekday, even if you spent more time in bed. This is because you are choosing to sleep through a time window when your biological clock favors REM sleep over NREM sleep and are close to a time window when your biological clock is on high alert for waking.

Now aside from keeping a habitual sleep schedule, there are ways to optimize the amount of restorative, NREM sleep through supplements and pharmaceuticals and then reap the benefits of growth hormone at the same time. If you have ever gone into a GNC, The Vitamin Shoppe, or the like, you will always encounter posters for a product that feature an insanely ripped—naturally or unnaturally—person and some marketing tagline like "proprietary protein blend." Jargon aside, what is in a "proprietary protein blend" and how can it benefit performance (and sleep)? First off, a "proprietary protein blend" contains all or some of the twenty common amino acids, which as you may remember from your sixth grade biology class, are the "building blocks of life." One of the more common amino acids found in these "proprietary protein blends" is L-arginine. The "L" refers to how the molecule structure of the protein is arranged, also known as a Fisher projection.

It just happens that L-arginine can increase the amount of deep, restorative sleep in both rats and men. Long-term intake of L-arginine in rats has been shown to increase deep, restorative NREM sleep during the night[8]. Most rodents are nocturnal, or night-active, so essentially L-arginine increased their napping. As I discussed in the previous chapter, napping does improve mental and motor performances in both rats and men. In human trials, arginine aspartate—a blend of L-arginine and another amino acid known as L-aspartate—increased deep NREM sleep, and most notably, increased the release of growth hormone induced by NREM sleep by 60 percent. One individual in this study was particularly sensitive to arginine aspartate and saw a 162 percent increase in NREM-stimulated release of growth hormone[9]. In the world of exercise physiology, L-arginine has also been shown to accelerate the healing process of an injured bone, muscle, or damaged tissue[10].

For athletes, L-arginine products are on the market and many are labeled as "banned substance free" when taken in the prescribed amounts. There are similar products on the market too, such as nitric oxide enhancers. Biochemically, L-arginine is used to make nitric oxide, which enhances vasodilation, thereby increasing blood flow to organs most needed for exercise like the brain, heart, and muscle. In fact, nitric oxide is the active ingredient of Viagra®; it was first used as a heart medication before it was serendipitously discovered to famously act elsewhere (side note: many of male teammates have taken nitric oxide enhances like N.O.-Xplode™ prior to a competition, and I never cease to joke with them that they better be careful for fear of me watching them slam a barbell kept close to the upper leg, groin, and lower chest area while performing an Olympic lift).

A last product aimed to increase the release of growth hormone that has received much media attention courtesy of an NFL superstar--Ray Lewis--is deer antler. Ray Lewis has bragged about the healing properties of a topical (spray) form of deer antler. Very recently, I have also become interested in deer antler since a bottle of velvet deer antler was given to me during the Reebok Crossfit Regionals by one of the on-site vendors. At the time, I had been dealing with excruciating rotator cuff tendonitis and was halfway through a weekend of workouts nicknamed "shoulder annihilation" by the commentators. Needless to say, I was willing to take any anti-inflammatory that was legal that weekend; my sponsor at the time, Maximum Human Performance, also gave me fish oil supplements, which have an abundance of omega fatty acids that have reliably been shown to reduce inflammation.

As for velvet deer antler, the owner of the company had taken velvet deer antler throughout his time with minor and major league baseball teams when he had dealt with worse injuries to his rotator cuff relative to me: tears and subsequent surgical repair. After confirming that velvet deer antler was legal to take like any responsible athlete, I used them for the remainder of the Regionals weekend and for weeks after the competition. Like the scientist that I am, I also made certain to do a literature search on a biomedical database for the impact that velvet deer antler has on physiology and sleep. There is not much.

In nature, velvet deer antler comes from the middle layer of the deer antler. Deer lose their antlers during some seasons and re-grow them during others. It is from this middle layer of the antler that the healing properties emerge. In brief, velvet deer antler stimulates the release of growth hormone. It has also been known to improve sleep in insomniacs, but its general effects on sleep in a healthy population or its widespread physiological effects remain to be investigated. To conclude, as goes with caffeine, it's perfectly fine to use the science of sleep and supplements for optimizing the release of growth hormone and subsequently improving sleep, just within reason.

Moving on, I am now going to talk about anabolic androgenic steroids (AAS), which I certainly do not approve the use of, but at the same, cannot ignore because they have a huge, huge, huge impact on the brain, physiology, and behavior. Steroids are powerful. This is probably an understatement. Why is this so? Hormones can act in one of two places in the cell—the cytoplasm or the nucleus. The effects of hormones in the cytoplasm are noteworthy because they can alter protein synthesis. But the effects of hormones like AAS in the nucleus are beastly because they can alter the force of life: DNA. Steroids can alter the central dogma of biology, which in classrooms refers to the conversion of DNA to RNA to protein. The central dogma of biology affects everything in the body and brain. Now that it is clear why AAS have such an immediate and significant effect the central dogma of biology, let's narrow our discussion to the effects of AAS on brain structure and function.

There are several AAS naturally produced by the body—testosterone, androstenedione, and dihydrotestosterone. As we know, testosterone is the hormone that defines a man, although women also release a fair amount. In fact, it's been shown that female athletes have higher levels of testosterone than their non-athlete peers. Whether or not more testosterone is a hard-wired, pre-destined trait of a female athlete or a byproduct of athletic training is an interesting topic of discussion. In the world of endocrinology, AAS and related reproductive steroids such as estrogen and progesterone can be used to "mascinulize" a female or "feminize" a male. This is routinely done in endocrine laboratories, including one that I work in to investigate sex differences in sleep in animal models of study.

"Mascinulizing" a female or "feminizing" a male can be accomplished in many ways. First, you can take out the essential reproductive organs. This was historically done in the Catholic Church to choirboys to prevent them from going through puberty, thereby preserving their melodic, high-pitched voices. "Feminizing" a male is accomplished through castration in rodent species too. From this, you may conclude that female is the default sex, which has been a topic of debate in the field of neuroendocrinology. In fact, I recently attended a meeting of the Society of Behavioral

Neuroendocrinology where the issue of females as the default sex was widely accepted although most scientists agree that it is an active versus passive process, requiring some sort of experimental manipulation.

Another means to alter the biological sex and stereotypical behaviors of a male or female rodent is to directly inject one of the sex steroids. The effect can be permanent if it is given soon after birth or before puberty. Scientists refer to this act of permanency as the critical period meaning that any sort of hormonal manipulation occurring beyond this critical period, depending on the type of manipulation is short-lived. The permanency that AAS and estrogen can have on a teenager's biological and skeletal makeup and athletic abilities seem to be well understood by gridiron coaches and parents.

I grew up in a region of the US known as the "Rustbelt." The "Rustbelt" is comprised of states and regions bordering the Ohio River and nearby Great Lakes, including Ohio, Pennsylvania, West Virginia, Indiana, Illinois, and Michigan. The "Rustbelt" once prospered in the steel and coal industries during times of war, but nowadays, is largely impoverished, suffers from diaspora (i.e. "white flight"), and struggles to find a new economy. Despite economic decline, many communities in the "Rustbelt" make large financial, political, and social investments in high school football programs.

On some occasions, the investment can be too overbearing and can result in the cover up of horrific crimes committed by a community's football team as was the case of the recent Steubenville rape trials[11], which was a rivalry school district of my own high school. I bring up the "Rustbelt's" significant investment in high school football programs to your attention because I have been well aware of high school football players abusing AAS partly out of encouragement of coaches and parents. In the US alone, it's estimated that anywhere from 6-11 percent of high school athletes take AAS with the majority of the offenders being high school football players.

The "benefit" of taking AAS in high school versus later in college, of which I use quotations for because I'm simply looking at it from the perspective of drug abusers, is that many school districts, conferences, and state tournament associations do not have drug-testing policies in place; I have competed and I have had athletes compete in the state track and field tournament for years and neither them nor me have even been selected for random drug testing after making it on the podium. What this means is that many high school (and even middle school) athletes can abuse AAS and reap long-term benefits in size and strength. The reason for this large effect of AAS on size and strength during puberty manifests from biological significance of puberty, itself.

Puberty is designed transform one's body into an adult because hormones re-wire the brain and body. Hormones, particularly the sex steroids, remarkably change one's structural makeup and affect one's behavior, as parents of teenagers know. Supplementation with AAS hormones can be additive, resulting in more *significant* change to the body, the brain, and behavior. Thus, for many high school athletes, one cycle of supplementation with AAS during pre-season training may be all they need to reach that "tipping point" in size and strength, of which athletes can maintain for a lifetime with continued training and competition.

So what effects do supplements containing AAS and sex steroids have on the brain? Unfortunately, most human-based studies focus on the detrimental and pathological effects of AAS on mental health: aggression, depression, and suicide. This is not the case in rodents. While you may be skeptical of the extrapolation of data collected in rodents to humans, you may be curious to know that many of the brain-altering effects of AAS and sex steroids occur in an area of the brain that is largely conserved across mammalian evolution: the hypothalamus.

As I have discussed, the hypothalamus is partitioned into many sub regions that are appropriately named for their location in space or function. One area of the hypothalamus that is markedly different between male and female rodents and is extremely sensitive to sex steroids is the sexually dimorphic nucleus of the preoptic area (SDN-POA). Many of these studies were done in the 1980s and 1990s, wherein scientists began to inject female hamsters with testosterone or males with estrogen. These scientists would also remove the hamsters' reproductive parts in order to see how the size of the SDN-POA as well as the cell types of the SDN-POA changed with adding, negating, or flip-flopping a sex steroid. The results were striking, markedly changing the masculinity and femininity of the SDN-POA through changes in its size and function or number of neurons communicating with other neurons.

Other areas of the hypothalamus that differ in size as a result of being biologically male or female include the ventromedial nucleus—which lies at the base (ventral) of the hypothalamus close to the midline of the brain—and the supraoptic nucleus which lies next to where the optic nerves cross. Within these areas of the hypothalamus, there are also differences in the amount of active neurons between males and females. Sex steroids like testosterone and estrogen also change the number of projections known as dendrites on neurons, which are essential for neuron-neuron communication. Moreover, changes in brain structure and function that have been induced by sex steroids in rodents have been correlated with changes in a range of behaviors like sexual promiscuity, general physical activity, and learning and memory. It's unfortunate that we do not have a corresponding breadth of knowledge in the human literature in regards to AAS and their impact on neuron structure in function.

"Masculinizing" or "feminizing" the brain and body can also be accomplished by altering how sex steroids bind to receptors in the cell—through a lock and key mechanism—once they are released. In fact, an alteration in the release and binding of sex steroids is the primary biological manifestation of gender identity disorders (GID), which have become a controversial subject in the world of sports. A few years ago, Ariel Levy wrote an informative piece for *The New Yorker* on the medical history and current state of GIDs in sports and society by focusing on a new-age running icon: Caster Semenya[12].

I have assigned this article to students in my vertebrate embryology and neuroscience courses since its publication, and I'm always amazed to hear such differing opinions on the "fate" of Caster Semenya from the student-athletes in the room. Caster Semenya is a professional middle-distance runner from South Africa who podiumed at the 2012 Olympics. She has also faced several bans from competition due to her gender being questioned. During this time, Caster was subjected to psychological, gynecological, and endocrine examinations to determine if she was male or female as it was suspected that she suffered from a GID known as androgen insensitivity syndrome (AIS). AIS is one of several GID wherein the person is ambiguously female or male, biologically and psychosocially.

Caster is not the first female to be suspected of having AIS or another GID though. We just know more about the biology of the condition than we did fifteen years ago; in the Atlanta 1996 Olympics, which was one of the last Olympic Games that subjected athletes to gender testing, eight female athletes were discovered to be genetically male and suffering from AIS[13]. This is at a rate much higher than the general population. So what is AIS?

As the name implies, the androgens released from the brain and testes have nowhere to bind. Since the androgen can't bind, the full blueprint of what makes a male characteristically male cannot come to fruition. The inability of androgens to bind to androgen-specific receptors results in a more masculine female or feminine male depending on your point of view. The tests for AIS are rigorous and not straightforward because the ability of androgens to bind to their receptors and the resulting anatomical and physiological characteristics can be full-blown or partial. In the case of full-blown AIS, a genetic test would reveal that this person is male, possessing one X and one Y chromosome. Medical imaging and gynecological tests would reveal external female-like genitalia at birth and across puberty as well as the presence of undeveloped testes that can cause great stomach pains inside the individual. Endocrine tests would reveal higher levels of testosterone.

To conclude, I want to refrain from delving into a discussion on the ethics and presence of GIDs on sports. Instead, I hope that you have been profoundly intrigued by the powerful effects that sex steroids, particularly AAS, can have on the brain, let alone the body. It would be nice for the sake of science if AAS were less stigmatized in society so that we could examine less clinically-orientated changes in the brain that result from AAS abuse such as the effects of AAS on attention, reaction time, and learning and memory as we have found in rodents. But, the rise in abuse of AAS in high school, collegiate, and professional sports coupled with harsher punishment will probably not allow for it.

Amped Up

There is no contention that competition is stressful for any athlete: amateur, collegiate, or professional. No matter how often I compete, I still have trouble sleeping the night before a competition. Friends of mine who play in the NFL or run professionally tell me the same. The difference between an amateur versus professional athlete though is that the professional thrives from the stress of a competition, particularly physiological stress, whereas an amateur chokes. I detailed the biology of choking in “The Athlete Brain in Competition” if you happened to have skipped ahead, but here is a refresher.

Aside from the psychological stress of competition, which can range from a fear of losing, disappointing teammates, or having a lack of self-confidence, the physiological stress of competition is driven by the hypothalamic-pituitary-adrenal (HPA) axis. This axis requires crosstalk between the brain, the adrenal glands, the muscles and other organs important for continued exercise. Activation of the HPA begins in an area of the hypothalamus known as the paraventricular nucleus (PVN) wherein corticotrophin-releasing hormone (CRH) is released. From the PVN, the HPA signal travels to the pituitary gland, which releases adrenocorticotropic hormone (ACTH). At last, cortisol is released from the adrenal glands, initiating a mobilization of energy stores for use.

Because of its importance in competition, the HPA axis could be a significant target for performance-enhancing companies and drug doping agencies. Many of the pre-workout shakes on the market have ingredients that “claim” to boost the HPA axis, particularly at the level of CRH in the brain. I say “claim” because the production of performance-enhancing drugs is largely unregulated by the government. Companies do not have to report their active ingredients and are not required to undertake scientific and clinical trials as I alluded to earlier in the chapter. Despite this lack of governmental foresight, there are prescription drugs on the market that do alter HPA axis activity of which I'll discuss in a bit.

Most studies that have studied differences in the HPA axis of trained athletes versus non-athletes have studied the endpoint of the axis—the release of cortisol from the adrenals—and have not focused on what's going in the brain at the level of the hypothalamus (CRH) and the pituitary gland (ACTH). This lack of knowledge is understandable because taking blood and urine samples which contain significant amounts of cortisol is certainly less invasive and risky than attempting to probe and measure the release of biological substances from the brain.

At any rate, the impact of the HPA axis on athletic performance has been widely studied in endurance athletes. The highest surges in cortisol release in distance runners occur during and after a training session or competition. Cortisol provides energy for the brain and muscles by commanding the conversion of glycogen to glucose in the liver as well the availability of fat. These energy reserves are then circulated to tissues in need. Cortisol also keeps inflammation and the pains of training at a manageable level. This is exactly why many athletes do not know the origins of their bumps and bruises or severities of their injuries post-competition. In college, my teammates and I referred to them as "party fouls" in an attempt to appear as cool as our fellow classmates who spent their weekends partying instead of traveling and competing. We were probably cooler and having more fun in hindsight.

Because the training for endurance-based sports is inherently longer, so is the release of cortisol. This doesn't seem to be a long-term health concern unless there is some aspect of overtraining. Now this is where cortisol research in lieu of athletic performance has received the most attention. Overtraining can be common to any training program. The point of training for anything competitive is to force the body to adapt to new physiological challenges and to recover quickly from these challenges, hoping to stay healthy and mobile. This is where having good genetics does help. At some point, training hard for three hours a day appears "normal" for the joints and brain. The body can withstand this type of high intensity training regimen for a few months, but anything past this time window can tip the scale and prevent any sort of recovery.

Overtraining wreaks havoc on joints, muscles, stomach, adrenals, and brain. I am well familiar with the symptoms of overtraining. Not only do you start counting down the days until the week of the competition where you essentially do not train, but workouts that normally do not tax the body cause you to puke and feel extremely achy. In the world of exercise physiology, overtraining is pinpointed by measuring cortisol release. Over-trained athletes who complain of symptoms like I have experienced often have persistently elevated and extremely high levels of cortisol. Thus, it's apparent that cortisol provides our bodies with organic protection from injury, but only to a certain extent.

Instead of focusing on the consequences of overtraining, how does a cortisol imbalance catalyzed by overtraining impact the brain? This has been widely studied within psychiatric communities dealing with anxiety and depression, but receives much less attention in athletic communities. Regardless, the effects of cortisol on brain anatomy and function are similar. The HPA axis is a classic textbook example of a negative feedback loop. A negative feedback loop refers to any physiological system wherein an intermediate and end product like ACTH and cortisol, respectively, can stop further release of the beginning product like CRH. Negative feedback loops do not just apply to hormones. There are plenty of negative feedback loops in the cell that control rates of gene and protein expression. Because the HPA axis is a negative feedback loop, this makes it permissible for cortisol to act in the brain in order to shut off the production of CRH and ACTH. This "go-stop" mechanism works very well under a normal lifestyle, but falters with overtraining or in someone experiencing intense anxiety because the brain can rapidly habituate to the binding of cortisol in the brain to shut off the production of CRH and ACTH. The brain's habituation to cortisol results in more cortisol secretion and a faulty "go-stop" mechanism, creating a downward spiral of effectiveness.

Although a faulty HPA axis has not been widely studied in the world of exercise physiology, it has been a major focus in labs that focus on the neuroscience of emotion. The two major areas of the brain that regulate our emotional stability or instability are the amygdala and the hippocampus, which neighbor the hypothalamus. The hippocampus is a very distinctive brain structure because it looks like a sea horse. In fact, the etymology of hippocampus derives from the Greek words *hippos* meaning horse and *kampos* meaning sea monster. Cortisol is capable of binding at the level of the hippocampus, which then shuts off the HPA axis.

However, high and persistent secretion of cortisol renders the hippocampus less sensitive and the hippocampus can't shut off the HPA axis as well. Further, the constant binding of cortisol in the hippocampus causes the hippocampus to degrade, killing neurons and connections, making the feedback loop even less effective. This has been has been demonstrated post-mortem in studies of anxiety and depression in rodents and humans. Although we know nothing about the origins of cortisol binding in the brain in response to overtraining, I do know from personal experience that an over-trained athlete can be quite irritable.

Because cortisol is a key biological factor that dictates the extent of recovery and one's training future, is there any way to trick the body into thinking that less cortisol was released? Yes, there is. It is called dexamethasone, which ironically enough, is commonly prescribed in drug rehabilitation programs to curb drug cravings. In the world of athletics, dexamethasone is commonly abused by distance runners and body builders. Dexamethasone suppresses HPA axis activity thereby suppressing the negative consequences of overtraining and allowing the body to recover quicker under a high intensity workload. The NCAA and other drug doping agencies are already ahead of the research though because dexamethasone is a banned substance.

Before I discuss the most exciting, next-generation of drug doping, at least to me, I do want to talk about some other drugs that get athletes amped up. Caffeine is not the only psychostimulant for rapidly increasing alertness and athletic performance. "Neo" psychostimulants like Ritalin and Adderall, which are overprescribed to children and are equally as abused by adults, are also a problem amongst athletes. A few years ago, a report came out in *The New York Times* about the abuse of these psychostimulants by NFL players in order to stay focused during the four hours of play on game day[14]. In fact, the abuse of Adderall by NFL players has increased by 75 percent within the past few years with several suspensions being instituted.

The paradox of Ritalin and Adderall is that they are used to calm the minds and bodies of people who are diagnosed with attention deficit disorder (ADD) or attention deficit hyperactivity disorder (ADHD). This knowledge can confer accurate diagnosis of ADD and ADHD. In people who do not have ADD and ADHD, Ritalin and Adderall can still give you "tunnel vision" and enable you to focus and perform well on difficult, menial, or cumbersome tasks for significant periods of times, sometimes for days, but it will also makes you extremely hyper.

I will admit that I have taken an Adderall pill during exam week as an undergraduate and graduate student whenever I had multiple exams in a day. The mind-altering effects and the end results in my scholastic performance were fantastic;

I never left the room with a foggy brain, and was excited, yes excited, to take the next exam. I would also have an amazing few days of track practice. Even if it was not a difficult workout that pushed me to a metabolic threshold, causing me to puke and my legs and posterior chain to burn, I would recover quickly and have plenty of energy on reserve before the next set of sprints. My body may have actually depleted itself of all its energy reserves by the end of those sprint intervals, but I certainly did not feel it.

I have searched for research on biomedical databases for how Adderall increases physical athletic performance. There is little. However, I did learn that Adderall is more widely abused by male athletes, and has high rates of abuse for sports like football, lacrosse, and wrestling. Female athletes also abuse Adderall, but the sport of play does not dictate the level of abuse for females unlike it does for males[15].

At the level of the brain, researchers have found the origin of Adderall's calming effect for the treatment of ADD and ADHD. The answer lies in calming the activity of neurons releasing dopamine. This study wasn't done in humans, but in a group of mice that had mental and behavioral states similar to individuals with ADHD[16]. These mice were ADHD because they lacked a transporter that allowed dopamine to be recovered after its release. Also, when these researchers gave these mice methamphetamines—which is the drug category that Adderall belongs to—their nerve cells that released dopamine were calmer and were not as hyperactive as in normal (wild-type) mice. It should be noted that Adderall and Ritalin increase dopamine production on a grandiose not minor scale, generating high levels of alertness and motivation. But, the vast recruitment of brain centers to release dopamine also makes those drugs extremely addictive.

Because Adderall and Ritalin has such a profound effect on dopamine, I can't help but to attribute the great training day that I experienced while on Adderall to changes in dopamine signaling; in "The Athlete Brain", I talked about how enhancements in dopamine communication may forestall physiological fatigue. To conclude, while the enhancement of physical athletic performance by psychostimulants such as cocaine, Adderall, and Ritalin is largely unknown, the NFL, NCAA, and other drug doping agencies do police for abused psychostimulants, even illicit ones like cocaine, because athletes are role models in many communities and so drug doping agencies attempt to uphold these societal expectations even if athletes do not.

New Frontiers in Drug Doping.

This is a frontier that I am excited, actually more ecstatic about, because my research area--behavioral genetics--will finally get the media attention that it deserves. In grade school, we learned that genes and complexes known as amino acids are the "building blocks of life." When these genes and amino acids make a mistake, we transform into mutants, for better or for worse. The impact that genes and amino acids have on our lives may not be apparent in a healthy person, but their impact is readily noticeable by someone who suffers from a birth defect or cancer. Both birth defects and cancer result from an error in the replication of genes and coding for amino acids. But what if there was a mistake in the replication of genes and coding for amino acids that was passed down for generations and created a tribe of super athletes? Would these mutations make these super athletes ineligible for competition under the NCAA and other drug doping agencies? Further, what if you knew of a means to manipulate the reading of genes and encoding of amino acids that bestowed super athleticism?

In the last chapter on sleep, circadian rhythms, and athletic performance, I talked about two specific mutations known as single nucleotide polymorphisms (SNPs) that make mice and men be extreme "larks"—the gene *Per2*—or survive on 2-4 hours of sleep—the gene *Dec2.* These extreme traits were not just present in a single individual, but in many. Therefore, the discovery of a specific mutation in a gene that could make for super athletes would be no different than phenotypes manifest from SNPs of *Per2* or *Dec2* in that athletic-conferring SNPs would likely be found in many people and not just one unique individual.

If there happened to be a SNP or series of SNPs for athleticism, it would make logical sense for a professional or collegiate team or even a military branch to capitalize on these SNPs and screen and recruit people with them; a proposition of similar sorts was brought to my attention during my dissertation defense. I recommended that hospital and corporate boards recruit populations of shift workers with a *Per2* SNP because these employees are more capable of handling night and swing shifts since these shifts coincide with these individuals' "normal" sleep-wake schedules. Such a decision could also reduce on-the-job accidents and other related-expenses, and from the perspective of labor unions, make for better health and welfare.

If it hypothetically turned out that coaches and sporting franchises were screening for genes, would this be considered illegal by governing organizations? It is a more science-based approach towards recruiting much like Billy Bean's mathematical-based approach when he was manager of the Oakland A's, right? I will let you decide, but there are laboratories around the world that have found genes encoding for super athleticism in mice and humans. Most of the human work is documented in David Epstein's *The Sporting Gene* so I will focus on the species that I know most about: mice. About a year ago, I learned about a population of mice who behave very much like the Tarahumara tribe of ultramarathon runners of the Yucatan Peninsula in Mexico who were featured in Christopher McDougall's book, *Born to Run: A Hidden Tribe, Super Athletes, and the Greatest Race the World Has Never Seen.* These mice were genetically engineered at Case Western University in Cleveland, OH to have ten-times more PEPCK—a metabolic enzyme—in their muscles[17]. A genetic manipulation of such a "trivial" extent can't have a too big of an impact on general activity, exercise, and the recruitment of fuel sources, right?

First, these mutant mice were hyperactive. Analyses of how often these mice roamed around their cages revealed that they roamed seven- to ten-times more often than the normal mice. They were also hotheads or were more aggressive. When put on a treadmill—which is actually possible to do in rodents as I did this in graduate school with hamsters—the mice ran an average of 5,000 kilometers, or 3.1 miles at 20 meters/min at a single period of time. This means that these mutant mice ran nonstop for 4 hours whereas the normal mice only ran 200 meters at a time. There is a fantastic YouTube video produced by this research group that highlights how these mutant mice ran on a treadmill in comparison to the normal mice (http://www.youtube.com/watch?v=4PXC_mctsgY). It is remarkable.

One reason why these mutant mice were capable of effortlessly running ultra-marathons on a daily basis is that they were extremely efficient at using fats as a fuel source. They were also lighter, leaner, and ate more than the normal mice. Third, they produced less lactate, which any runner, swimmer, or Crossfitter knows, is the limiting factor for running, swimming, or Crossfitting longer, harder, and recovering quicker. To add more icing to the cake, these super athletic mice were also less susceptible to diabetes because they had lower circulating levels of insulin. Even better, they remained reproductively active longer and they lived longer.

Coaches and sporting organizations, I would start searching for these super PEPCK humans now. While I think that it would be unreasonable for the NCAA or any sporting organization to penalize coaches and general managers for using the next frontier in science and technology in order to capitalize on the genetically gifted athlete, enforcement of such sorts would not surprise me. Just think about how many hockey dads and soccer moms would find it unfair that the athletic future of their son or daughter was in the hands of their family's biological makeup.

If athletes do not happen to be born or hard-wired for a certain amount of athleticism that is controlled by their genetic makeup, would it still be possible to manipulate our genetic code after birth? Possibly. One of the intriguing topics in biomedical research today is the field of epigenetics; the idea that our environment, especially a stressful one, can directly alter the reading of our genes and coding for amino acids across childhood, adolescence, and adulthood. Usually the genetic modification comes in the form of adding a methyl group to the DNA, interfering with its encoding, or changing the rate at which the DNA is unwound from thread spool-like structures called chromatin.

I've seen tons of epigenetic data over the years at large professional conferences on circadian rhythms, sleep, reproductive biology, and drug addiction. I will say that the most bizarre and unusual epigenetic data that I encountered was at a meeting for the Society for Behavioral Neuroendocrinology. Here, Arthur Arnold—a pioneer in the study of hormones and behavior---presented data on male mice that had lower quality sperm only because their grandmother lived a stressful life. Unusual epigenetic stories in children of drug addicts have also been narrated on National

Public Radio shows like *Radiolab* and *Freakonomics,* but the impact of daily stress on the human genetic code, progeny, and subsequent athleticism is elusive.

To speculate, perhaps there is some biological truth to the declines in athletic performance that are often seen in kids of overbearing hockey dads and soccer moms as I discussed in "The Athlete Brain;" maybe the physiological and psychological stress of meeting coaches' and parents' expectations is too overwhelming for the minds and the genetic codes of these kids, leading to a downward spiral in athletic performance? I will admit that this specific hypothesis is highly implausible, but my point is that there may be some instances that remain to be documented in which an early life of stress (or lack thereof) affects one's genetic makeup in a way that it directly impacts exercise physiology.

There may also be a way to manipulate our genetic makeup for the enhancement of athletic performance by means of drugs. There is no question that these drugs could eventually be detected by drug doping agencies. In molecular biology, there are several ways that scientists can turn the reading of DNA and encoding of amino acids "on" and "off." One means is to give mice of a unique genetic makeup tetracycline—an antibiotic usually prescribed for the treatment of gonorrhea—to shut off the expression of specific genes. This happens because the tetracycline binds to a complex in the cell known as an operon complex where the gene is expressed.

The biological substance cyclohexamide—CHX—is another means to turn "off" genes at least in cell culture.

Obviously, both of these approaches--tetracycline and CHX--are used in a controlled laboratory setting, but what if genes that bestowed super athleticism could be turned "on" and "off" in a safe manner with few long-term side effects? Would athletes use them? Most likely, yes. And what would be the level of oversight adopted by drug doping agencies? It could be decades before such technology was discovered by drug doping agencies, but maybe I'm not giving drug-doping agencies enough credit. Maybe drug-doping agencies have already forecasted this mode of PED abuse, although I can find no evidence for it.

As I conclude this chapter, I hope that you have been well educated about the many classes and forms of drug doping. One's intent for drug doping is not about getting muscles bigger and recovering quicker. It is also about keeping the mind focused and reaction time low. PEDs affect the brain on many levels by exploiting the release of certain neurochemicals such as dopamine, which can reduce mental and physical fatigue, or certain hormones like testosterone and growth hormone. While there may be options for improving athletic performance in the absence of drugs, such as getting enough deep, restorative sleep to optimize the release of growth hormone, people also want quicker results. While I'm not advocating for PED abuse, I will advocate for screening for genes that confer super athleticism. If not for improving the team's winning record, the screening for "athletic" genes will also benefit our scientific understanding of the athlete mind and body.

Epilogue

As I stated in the prologue, I am the first to admit that the broadcasting of research to the general public can be convoluted. Research articles are littered with jargon and complex graphs that are, in turn, often misinterpreted by the media. The bottom line is that I wrote this book with the intent of making the science relatable to you—the general public—by bridging the world of biomedical science with my broad athletic pursuits and those of notable athletes. By reading this book, I hope that you have gained an appreciation for the benefit of exercise for the brain and mental health, which is often overlooked by the "war on obesity" campaign.

Beyond the benefits of exercise for mental health, I hope that you have become fascinated by how plastic the brain is. I have demonstrated that the brain is able to respond to new environments and stress remarkably well and that there can be near permanent changes in how the brain re-wires itself and communicates with the rest of the body all with a little practice. This doesn't just happen as a budding adolescent (although it helps), but it persists through adulthood! For those of you who feel as though you are destined to live the life of a couch potato who can only watch others bask in athletic glory, I hope that you have come to realize that your brain and health can be quickly remedied.

In writing this book, there were three demographics that I sought to target: the exercise enthusiast—someone who leisurely works out for personal reasons whether it's for the love of fitness, vanity, general health and well-being; the competitive athlete—or someone who *likes* to push themselves to new physiological and mental limits, goes above and beyond what their body can handle, and does this type of regimen day in, day out, rain or shine; and the sports junkie—someone who has profound statistical and biographical knowledge of many sports, one sport, or a particularly team and has no issue about using their disposable income in support of their favorite sports organization or team. I hope that those of you who fall into one of these three demographics found knowledge and amusement in most, if not all of the chapters.

For the exercise enthusiast, you should feel empowered by the widespread changes in the brain that come with daily physical activity, ranging from the improvement of mood, memory, learning, alertness, coordination, and reaction time. I also expect that many of you will pay more attention to how much you sleep, when you sleep, and your quality of sleep.

For the sports junkie, I hope that you found amusement in the underlying biological makeup of athletes, sporting events, and play-by-plays. In "The Athlete Brain," I hope that you appreciated the recruitment of loyal and knowledgeable fans much like yourself to study where intense emotions experienced during a high-stakes game are regulated in the brain. For my competitive athletes, I hope that you can view this as a handbook for gaining a competitive edge as you continue your athletic endeavors. You have spent years training your brain and body to react and adapt to physiological and mental stress. You deserve a boost.

For young athletes, I hope that you find motivation from knowing the importance of practice for the brain's sake and how such re-wiring of the brain with practice can revolutionize athletic performance. For high school and college athletes, you can certainly use this book as empirical support in rebuttal to one of your non-athletic college peers who calls you some derogatory term like "dumb jock" or "meathead." Much like the exercise enthusiast, I hope that you begin to pay attention to your sleep schedule, especially during game day travel, and realize how much and when you sleep can dramatically affect athletic and scholastic performances. I hope that you find solace in the discussion of pre-competition butterflies and realize that it is perfectly natural, and remember not to think too much during a competition and rely instead on your muscle memory. Your brain knows what it has to do from the 10,000 hours of practice. Let it do it. Finally, I hope that you can take advantage of next-generation performance enhancers provided by science. Of course this requires careful research as to what you are exactly putting into your body and use in moderation, but there's nothing wrong with using science to gain a competitive edge if it is within the guidelines. To conclude, I have tremendously enjoyed the research and writing of this book. It's not every day that I can relate my years of blood, sweat, and tears on the track and in the gym to scientific concepts or have the opportunity to express them to the public rather than my own specialized scientific community.

Acknowledgments

First, I must thank the wonderful staff of WestBow Press who has patiently worked with me the past few years. Second, none of this material would have come to fruition without the many educators who I have had in neuroscience and physiology in college and graduate school. Your service to higher education is much appreciated. Third, the content of this book would have never come to fruition without the many coaches, trainers, and teammates who I have worked and competed with across the past two decades. Thank you for continuing to keep me motivated, healthy, and finding my inner self with athletic competition. A special thanks is due to fellow Spartan track and field alumna Adriane Blewitt who wrote the foreword. I would also like to thank TJ Murphy and Rachel Herz, authors of *Inside the Box: How Crossfit Shredded the Rules, Stripped Down the Gym, and Rebuilt My Body* and *The Scent of Desire: Discovering Our Enigmatic Sense of Smell,* respectively, for navigating me through the publishing of non-fictional science writing. Lastly, I would like to thank family and friends who have supported my journey to broadcast this idea that I brainstormed many years ago into a book.

Endnotes

Barbells and Brains

1. Greenwood BN, Foley TE et al. (2011) Long-term voluntary wheel running is rewarding and produces plasticity in the mesolimbic reward pathway. *Behavioural Brain Research* 217: 354-62.
2. Lukaszyk A, Buczko W et al. (1983) The effect of strenuous exercise on the reactivity of the central dopaminergic system in the rat. *Polish Journal of Pharmacology and Pharmacy* 35: 29-36.
3. Smith MA, Schmidt KT et al. (2008) Aerobic exercise decreases the positive-reinforcing effects of cocaine. *Drug and Alcohol Dependence* 98: 129-35.
4. Webb IC, Baltazar RM et al. (2009) Diurnal variations in natural and drug reward, mesolimbic tyrosine hydroxylase, and clock gene expression in the male rat. *Journal of Biological Rhythms* 24(6): 465-476.
5. Ehringer MA, Hoft NR et al. (2009) Reduced alcohol consumption in mice with access to a running wheel. Alcohol 43: 443-52.
6. Collingwood TR, Sunderlin J et al. (2000) Physical training as a substance abuse prevention intervention for youth. *Journal of Drug Education* 30:435-45.
7. Mistlberger RE, Antle MC et al. (2000). Behavioral and serotonergic regulation of circadian rhythms. *Biological Rhythms Research* 31(3): 240-283.

[8] McDuff DR, Baron D (2005). Substance abuse in athletics: a sports psychiatry perspective. *Clinics in Sports Medicine* 885-897.

[9] Kinnaman KA, Mannix RC et al. (2013). Management strategies and medication use for treating paediatric patients with concussions. *Acta Paediatrics* 102: e424-428.

[10] Brager AB, Ehlen JC et al. (2013). Sleep loss and the inflammatory response in mice under chronic environmental circadian disruption. *PLoS One* 8(5): e63752.

[11] Ostrowski K, Rohde T et al. (1999). Pro- and anti-inflammatory cytokine balance in strenuous exercise in humans. *Journal of Physiology* 515: 287-291.

[12] Castanon-Cervantes O, Wu M et al. (2010). Dysregulation of inflammatory responses by chronic circadian disruption. *Journal of Immunology* 185: 5796-5805.

[13] Cotman CW, Berchtold NC et al. (2002). Exercise: a behavioral intervention to enhance brain health and plasticity. *Trends in Neuroscience* 25(6): 295-301.

[14] Goldin P, Ziv M et al. (2013). MBSR vs aerobic exercise in social anxiety: fMRI of emotional regulation of negative self-beliefs. *Social Cognitive and Affective Neuroscience*. 8:1 (65-72).

[15] Dubreucq S, Marsicano G, Chaouloff F (2011). Emotional consequences of wheel running in mice: what are the appropriate controls? *Hippocampus* 21(3): 239-242.

[16] Hofstetter JR, Mayeda AR et al. (1995). Quantitative trait loci (QTL) for circadian rhythms of locomotor activity in mice. *Behavioral Genetics* 25(6): 545-56.

[17] De Moor MH, Liu Y et al. (2009). Genome-wide association study of exercise behavior in Dutch and American adults. *Medicine and Science in Sports and Exercise*. 2009;41(10):1887–95.

[18] Brager AB, Yang T et al. (2014). Differential changes in NREM sleep amounts and stroke outcome in pre-conditioned mice: Influences of *Bmal1*. *Sleep* 6(Suppl): A29.

The Athlete Brain

1. Kanters MA, Boccaro J et al. (2008). Supported or pressured? An axamination of agreement among parent's and children on parent's role in youth sports. *Journal of Sports Behavior* 31: 64-80.
2. www.ncaapublications.org
3. Alilian WJ, Li X et al. (2008). Light-induced rescue of breathing after spinal cord injury. *Journal of Neuroscience* 28: 11862-70.
4. Volkmann J, Schnitzler A et al. (1998). Handedness and asymmetry of hand representation in human motor cortex. *Journal of Neurophysiology* 79: 2149-54.
5. McGrath, B (Dec 31 2011). Does football have a future? *The New Yorker.*
6. Bears, Connors, and Paradiso. *Introduction to Neuroscience.* 2nd edition.
7. Zimmer, Carl (Apr 2010). The brain: Why athletes are geniuses. *Discover Magazine.*
8. Hernandez, Michelle (Nov 13 2011). Athletes are the problem. *The New York Times* (Opinion Pages).
9. Fortune 500's baller and shot caller CEOs (2011). CNN Money.
10. Di X, Zhu S et al. (2012). Altered resting brain function and structure in professional badminton players. *Brain Connectivity* 2: 225-33.
11. Abreau AM, Macaluso E et al. (2012). Action anticipation beyond the action observation network: a functional magnetic resonance imaging study in expert basketball players. *European Journal of Neuroscience* 35: 1646-54.
12. Wei G, Zhang Y et al. (2011). Increased cortical thickness in sports experts: a comparison of diving players with the controls. *PLoS One* 16: e17112.
13. Alves H, Voss MW et al. (in press). Perceptual-cognitive expertise in elite volleyball players. *Frontiers in Psychology.*

[14] Del Percio C, Babiloni C et al. (2009). "Neural efficiency" of athletes" brain for upright standing: a high-resolution EEG study. *Brain Research Bulletin* 79: 193-200/

[15] Fogel SM, Smith CT (2011). The function of the sleep spindle: a physiological index of intelligence and a mechanism for sleep-dependent memory consolidation. *Neuroscience & Biobehavioral Reviews* 35: 1154-1165.

[16] Ajemian R, D'Austilio A et al. (2010). Why professional athletes need a prolonged period of warm-up and other peculiarities of human motor learning. *Journal of Motor Behavior* 42: 381-388.

[17] Kringelbach ML, Jenkinson N, et al. (2007). Translational principles of deep brain stimulation. *Nature Reviews Neuroscience*. 8:623–635

[18] Reis J, Schambra HM et al. (2009). Noninvasive cortical stimulation enhances motor skill acquisition over multiple days through an effect of consolidation. *Proceedings of the National Academies of Sciences of the United States of America* 106: 1590-1595.

[19] Botzung A, Rubin D et al. (2010). Mental hoop diaries: emotional memories of a college basketball game in rival fans. *Journal of Neuroscience* 30: 2130-2137.

[20] Gibson, O (6 Dec 2012). Britain's Olympic success leads to record boost in sports participation. *The Guardian UK*.

[21] (24 Jul 2010). Soccer's growth in the US seems steady. *The New York Times* (Sports Section).

[22] Pierce RC, Kumaresan V (2006). The mesolimbic dopamine system: the final common pathway for the reinforcing effect of drugs of abuse? *Neuroscience & Biobehavioral Reviews* 30: 215-238.

[23] Cohen JY, Haesler S et al. Neuron-type specific signals for reward and punishment in the ventral tegmental area. *Nature* 482: 85-88.

[24] Greenwood BN, Foley TE et al. (2011). Long-term voluntary wheel running is rewarding and produces plasticity in the

mesolimbic reward pathway. *Behavioral Brain Research* 217: 354-362.

[25] Treadway MT, Buckholtz JW et al. (2011). Dopaminergic mechanisms of individual differences in human effort-based decision-making. *Journal of Neuroscience* 31: 6170-6176.

[26] Gilbert C (1995). Optimal physical performance in athletes: key roles of dopamine in a specific neurotransmitter/hormonal mechanism. *Mechanisms of Ageing and Development* 84: 83-102.

[27] Lieberman D, Venkadesan M et al. (2010). Foot strike patterns and collision forces in habitually barefoot versus shod runners. *Nature* 463: 531–535.

[28] Sternberg WF, Bailin D et al. (1998). Competition alters the perception of noxious stimuli in male and female athletes. *Pain* 76: 231-238.

[29] De Crée C (1990). The possible involvement of endogenous opioid peptides and catecholestrogens in provoking menstrual irregularities in women athletes. *International Journal of Sports Medicine* 11: 329-348.

The Athlete Brain in Competition

[1] Peterson SL, Weber JC, and Trousdale WW (1967). Personality traits women in team sports vs women in individual sports. *Research Quarterly* 38(4): 686-690.

[2] Malumphy TM (1968). Personality of women athletes in intercollegiate competition. *Research Quarterly* 39(3): 610-620.

[3] Goldstein AN, Greer SM et al. (2013). Tired and apprehensive: anxiety amplifies the impact of sleep loss on aversive brain anticipation. *Journal of Neuroscience 33* (26), 10607-154.

[4] Rice TB, Dunn RE et al. (2010). Sleep-disordered breathing in the National Football League. *Sleep* 33(6): 819-824.

[5] Thomson CJ, Hanna CW et al. (2012). The -521 C/T variant in the dopamine-4-receptor gene (DRD4) is associated with skiing

and snowboarding behavior. *Scandinavian Journal of Medicine and Science in Sports* 23(2): e108-113.

6 Eilam D, Izhar R, Mort J (2011). Threat detection: Behavioral practices in animals and humans. *Neuroscience & Biobehavioral Reviews* 35 (4): 999

7 Kurkjian Tim. It's Friday the 13th. Be afraid amused. *ESPN: The Magazine*. September 13, 2013.

8 Chang Y, Lee JJ et al. (2011). Neural correlates of motor imagery for elite archers. *NMR in Biomedicine* 24(4): 366-372.

9 Brager, Allison. Ohio and Charles Nagy Syndrome. *Daily Kent Stater*. January 15, 2008.

10 Sian Beilock. *Choke: What the Secrets of the Brain Reveal About Getting It Right When You Have To.* Atria Books, August 9, 2011.

Eat, Train, Sleep

1 Dibner C, Schibler U et al. (2010). The mammalian circadian timing system: organization and coordination of central and peripheral clocks. *Annual Review of Physiology* 72: 517-549

2 O'Neill JS, Reddy AB (2011). Circadian clocks in human red blood cells. *Nature* 469: 498-503.

3 Moore, RY (1997) Circadian Rhythms: Basic neurobiology and clinical applications. *Annual Review of Medicine* 48: 253-266.

4 Hammer SB, Ruby CL et al. (2011) Environmental modulation of alcohol intake in hamsters: effects of wheel running and constant light exposure. *Alcoholism: Clinical and Experimental Research* 24: 1651-1658.

5 Mistlberger RE, Antle MC et al. (2000) Behavioral and serotonergic regulation of circadian rhythms. *Biological Rhythms Research* 31: 240-283.

6 Glass JD, Guinn J et al. (2010) On the intrinsic regulation of neuropeptide Y release in the mammalian suprachiasmatic nucleus circadian clock. *European Journal of Neuroscience* 31:1117-1126.

[7] Shea S, Hilton M et al. (2011) Existence of an endogenous circadian blood pressure rhythm in humans that peaks in the evening. *Circulation Research* 108: 980-984.

[8] Czeisler C, Gooley JJ (2007). Sleep and circadian rhythms in humans. *Cold Spring Harbor Symposium in Quantitative Biology* 72: 579-597.

[9] Tarokh L, Raffray T et al. (2010). Physiology of normal sleep in adolescents. *Adolescent Medicine: State of the Art Reviews* 21:401-417.

[10] Wolff G, Esser KA (2012). Schedule exercise phase shifts the circadian clock in skeletal muscle. *Medicine and Science in Sports and Exercise* 44: 1663-1670.

[11] Brager AB, Prosser RA et al. (2011). Circadian and acamprosate modulation of elevated ethanol drinking in mPer2 clock gene mutant mice. *Chronobiology International* 28: 664-672.

[12] Glass JD, Brager AB et al. (2012). Cocaine modulates pathways for photic and nonphotic entrainment of the mammalian SCN circadian clock. *American Journal of Physiology: Regulatory, Integrative, and Comparative Physiology* 302: R740-750.

[13] Spanagel R, Pendyala G et al. (2005). The clock gene Per2 influences the glutamatergic system and modulates alcohol consumption. *Nature Medicine* 11: 35-42.

[14] Xu Y, Toh K et al. (2007). Modeling of a human circadian mutation yields insights into clock regulation by PER2. *Cell* 128(1): 59.

[15] Carskadon MA, Acedbo C (2002). Regulation of sleepiness in adolescents: updates, insights, and speculation. *Sleep* 25(6): 606-614.

[16] http://www.sleepfoundation.org/article/hot-topics/backgrounder-later-school-start-times

[17] Atkinson G, Jones H et al. (2005). Effects of daytime ingestion of melatonin on short-term athletic performance. *Ergonomics* 48(11-14): 1512-1522.

[18] Van Dongen HPA, Dinges DF. Circadian rhythms in fatigue, alertness, and performance. In: Kryger MH, Roth T, and Dement

WC (Eds.), *Principles and Practice of Sleep Medicine* (3rd ed.): 391–399. W. B. Saunders, Philadelphia, Pennsylvania.

19 Srividya R, Mallick HM, Kumar VM (2006). Differences in the effects of medial and lateral preoptic lesions on thermoregulation and sleep in rats. *Neuroscience* 139(3): 853-864.

20 Corey TP, Shoup-Knox ML et al. (2011). Changes in physiology before, during, and after yawning. *Frontiers in Evolutionary Neuroscience* 3:7.

21 Tam J, Cinar R et al. (2012). Peripheral cannabinoid-1 receptor inverse agonism reduces obesity by reversing leptin resistance. Cell Metabolism 16: 167-179.

22 He Y, Jones CR et al. (2009). The transcriptional repressor DEC2 regulates sleep length in mammals. *Science* 325: 866-870.

23 Grandner MA, Jackson N (2013). Dietary nutrients associated with short and long sleep duration. Data from a nationally representative sample. *Appetite* 64: 71-80.

24 Cromie, W. Just sleep on it: and empty the brain's 'in box.' *Harvard Gazette* October 26, 2000.

25 Rasch B, Gais S, Born J (2009). Impaired off-line consolidation of motor memories after combined blockade of cholinergic receptors during REM sleep-rich sleep. *Neuropsychopharmacology* 34: 1843-1853.

26 Tortonese DJ, Short RV (2012). Biological rhythms, jet lag, and performance in Thoroughbred racehorses. *Equine Veterinary Journal* 44: 377.

27 Brager AJ, Ehlen JC et al. (2013). Sleep loss and the inflammatory response in mice under chronic environmental circadian disruption. *PLoS One* 8: e63752.

28 Holmes, Baxter. NBA players are losing sleep over this season. *Los Angeles Times* April 16, 2012.

29 Worthen JB, Wade CE (1999). Direction of travel and visiting team athletic performance: support for circadian dysrhythmia hypothesis. *Journal of Sport Behavior* 22: 279.

[30] Howatson G, Bell PG et al. (2012). Effect of tart cherry juice (Prunus cerasus) on melatonin levels and enhanced sleep quality. *European Journal of Nutrition* 51: 909-916.

[31] Pigeon WR, Carr M et al. (2010). Effects of tart cherry juice beverage on sleep of older adults with insomnia: a pilot study. *Journal of Medicinal Food* 13: 579-583.

[32] Zhao J, Tian Y et al. (2012). Red light and the sleep quality and endurance performance of Chinese female basketball players. *Journal of Athletic Training* 47: 673-678.

Central Doping

[1] Giang, Vivian. Here are the professions that drink the most coffee. *Business Insider*, Aug 28, 2012.

[2] Mora-Rodríguez R, García Pallarés J et al (2012). Caffeine ingestion reverses the circadian rhythm effects on neuromuscular performance in highly resistance-trained men. *PLoS One, 7* (4) e33807.

[3] Sinton CM, Petitjean CF (1989). The influence of chronic caffeine administration on sleep parameters in the cat. *Pharmacology Biochemistry and Behavior* 32(2): 459-462.

[4] Gandevia SC and Taylor JL (2006). Supraspinal fatigue: the effects of caffeine on human muscle performance. *Journal of Applied Physiology* 100: R1749-1750.

[5] Imagawa TF, Hirano I et al. (2009). Caffeine and taurine enhance endurance performance. *International Journal of Sports Medicine* 30(7): 485-488.

[6] Sanz MT, Rochiccioli P (1985). Comparison of spontaneous growth hormone secretion during the daytime and sleep in children with short stature. *Hormone Research* 22(1-2): 17-23.

[7] Horne JA (1981). The effects of exercise upon sleep: a critical review. *Biological Psychology* 12(4): 241-290.

[8] Monti JM, Jantos H (2004). Effects of L-arginine and SIN-1 on sleep and waking in the rat during both phases of the light-dark cycle. *Life Sciences* 75(17): 2027-2034.

[9] Besset A, Bonardet A et al. (1982). Increase in sleep related GH and Prl secretion after chronic arginine aspartate administration in man. *Acta Endocrinologica* 99(1): 18-23.

[10] Stechmiller JK, Childress B et al. (2005). Arginine supplementation and would healing. *Nutrition & Dietetics* 20: 52-61.

[11] Levy, Ariel. Trial by Twitter. *The New Yorker.* August 5, 2013.

[12] Levy, Ariel. Either/Or. *The New Yorker.* November 30, 2009.

[13] Harrell, Eben. Is a female track start a man? No simple answer. *Time*. August 25th, 2009.

[14] Battista, Judy. Drug Focus Is at Center of Suspensions. *The New York Times.* December 1, 2012.

[15] Veliz P, Boyd C et al. (2013). Adolescent athletic participation and nonmedical Adderall use: an exploratory analysis of a performance-enhancing drug. *Journal of Studies on Alcohol and Drugs* 74(5).

[16] Francesco N, Bonito-Oliva A et al (2010). Role of aberrant striatal dopamine D1 receptor/cAMP/protein kinase A/DARPP32 signaling in the paradoxical calming effect of amphetamine. *Journal of Neuroscience* 30(33): 11043-11056.

[17] Hanson RW, Hakimi P (2008). Born to run: the story of the PEPCK-mus mouse. *Biochimie* 90(6): 838-842.

Index

A

B

C

D

E

F

G

H

I

J

L

M

N

O

P

Q

R

S

T

U

V